Lecture Notes in Geoinformation and Cartography

Series editors

William Cartwright, Melbourne, Australia
Georg Gartner, Wien, Austria
Liqiu Meng, München, Germany
Michael P. Peterson, Omaha, USA

About the Series

The Lecture Notes in Geoinformation and Cartography series provides a contemporary view of current research and development in Geoinformation and Cartography, including GIS and Geographic Information Science. Publications with associated electronic media examine areas of development and current technology. Editors from multiple continents, in association with national and international organizations and societies bring together the most comprehensive forum for Geoinformation and Cartography.

The scope of Lecture Notes in Geoinformation and Cartography spans the range of interdisciplinary topics in a variety of research and application fields. The type of material published traditionally includes:

- proceedings that are peer-reviewed and published in association with a conference;
- post-proceedings consisting of thoroughly revised final papers; and
- research monographs that may be based on individual research projects.

The Lecture Notes in Geoinformation and Cartography series also includes various other publications, including:

- tutorials or collections of lectures for advanced courses;
- contemporary surveys that offer an objective summary of a current topic of interest; and
- emerging areas of research directed at a broad community of practitioners.

More information about this series at http://www.springer.com/series/7418

Georg Gartner · Haosheng Huang
Editors

Progress in Location-Based Services 2014

Springer

Editors
Georg Gartner
Haosheng Huang
Department of Geodesy and Geoinformation
Vienna University of Technology
Vienna
Austria

ISSN 1863-2246 ISSN 1863-2351 (electronic)
ISBN 978-3-319-11878-9 ISBN 978-3-319-11879-6 (eBook)
DOI 10.1007/978-3-319-11879-6

Library of Congress Control Number: 2014953956

Springer Cham Heidelberg New York Dordrecht London

Printed on acid-free paper

Springer is part of Springer Science+Business Media (www.springer.com)

Preface

Recent years have witnessed rapid advances in location-based services (LBS) with the continuous evolvement of mobile devices and communication technologies. LBS have become more and more popular not only in citywide outdoor environments, but also in shopping malls, museums, and many other indoor environments. They have been applied for emergency services, tourism services, intelligent transport services, gaming, assistive services, etc.

This book provides a general picture of recent research activities related to this field. Such activities emerged in the last years, especially concerning issues of outdoor/indoor positioning, smart environment, spatial modeling, personalization, context awareness, cartographic communication, novel user interfaces, crowdsourcing, social media, big data analysis, usability, and privacy. The innovative and contemporary character of these topics has led to a great variety of interdisciplinary research and studies, from academia to business, from computer science to geodesy.

The contributions in this book are a selection of peer-reviewed full papers submitted to the 11th International Symposium on Location-Based Services in Vienna (Austria) in November 2014, organized by the Research Group Cartography, Vienna University of Technology. We are grateful to all colleagues who helped with their critical reviews. Please find a list of their names in the "Reviewers" section.

The conference series on LBS has been held at

- 2002—Vienna, Austria
- 2004—Vienna, Austria
- 2005—Vienna, Austria
- 2007—Hong Kong, China
- 2008—Salzburg, Austria
- 2009—Nottingham, UK
- 2010—Guangzhou, China
- 2011—Vienna, Austria

- 2012—Munich, Germany
- 2013—Shanghai, China
- 2014—Vienna, Austria

The conferences themselves were a response to an increased interest in providing anyone, anything, anytime, and anywhere services. These conferences together offer a general overview of how LBS-related research has been evolving in the last years. The contributions of this book reflect the recent main areas of interest, including wayfinding and navigation, outdoor and indoor positioning, spatial-temporal data processing and analysis, usability, and application development.

We would like to thank our colleagues Manuela Schmidt, Felix Ortag, Florian Ledermann, and Günther Retscher for their help during the production of this book.

Vienna, Austria, September 2014

Georg Gartner
Haosheng Huang

Contents

Part I Wayfinding and Navigation

Is OSM Good Enough for Vehicle Routing? A Study Comparing Street Networks in Vienna 3
Anita Graser, Markus Straub and Melitta Dragaschnig

Calculating Route Probability from Uncertain Origins to a Destination 19
Carolin von Groote-Bidlingmaier, David Jonietz and Sabine Timpf

Visualization and Communication of Indoor Routing Information 33
Jukka M. Krisp, Mathias Jahnke, Hao Lyu and Florian Fackler

A Computational Method for Indoor Landmark Extraction 45
Hao Lyu, Zhonghai Yu and Liqiu Meng

Part II Positioning

On the Feasibility of Using Two Mobile Phones and WLAN Signal to Detect Co-Location of Two Users for Epidemic Prediction 63
Khuong An Nguyen, Zhiyuan Luo and Chris Watkins

3D Indoor Location on Mobile Phones Using Embedded Sensors and Close-Range Photogrammetry 79
Xiujuan Li, Yan Zhou and Hanjiang Xiong

Range Domain IMM Filtering with Additional Signal Attenuation Error Mitigation of Individual Channels for WLAN RSSI-Based Position-Tracking 91
Seong Yun Cho

Application of the Inertial Navigation System 3D-Self-Calibration-Method for the Minimization of the Measurement Uncertainty 105
Enrico Köppe, Daniel Augustin, Tabea Wilk, Andreas Subaric-Leitis, Achim Liers and Jochen Schiller

Part III Spatial-Temporal Data Processing and Analysis

Feature Selection in Conditional Random Fields for Map Matching of GPS Trajectories 121
Jian Yang and Liqiu Meng

Road Network Conflation: An Iterative Hierarchical Approach 137
Andreas Hackeloeer, Klaas Klasing, Jukka Matthias Krisp and Liqiu Meng

Analysing the Usage of Spatial Prepositions in Short Messages 153
André Dittrich, Daniela Richter and Christian Lucas

Using Location-Based Social Media for Ranking Individual Familiarity with Places: A Case Study with Foursquare Check-in Data 171
Wangshu Wang

Part IV Innovative LBS Applications

A Space Time Alarm 187
Adrian C. Prelipcean, Falko Schmid and Takeshi Shirabe

Urban Emotions—Geo-Semantic Emotion Extraction from Technical Sensors, Human Sensors and Crowdsourced Data 199
Bernd Resch, Anja Summa, Günther Sagl, Peter Zeile and Jan-Philipp Exner

Citizens as Expert Sensors: One Step Up on the VGI Ladder 213
Farid Karimipour and Omid Azari

LBS-Based Dilemma Zone Warning System at Signalized Intersection 223
Yi Li, Junhua Wang and Lanfang Zhang

ATSSS: An Active Traffic Safety Service System in Pudong New District, Shanghai, China 239
Hangbin Wu, Chun Liu, Junhua Wang, Lianbi Yao, Shuhang Zhang, Yi Li, Zhengning Li, Cheng Liu and Shouen Fang

Part V General Aspects of LBS

Bridging the Gap Between Field- and Lab-Based User Studies for Location-Based Services 257
Ioannis Delikostidis, Holger Fritze, Thore Fechner and Christian Kray

Challenges of Location-Based Services Market Analysis: Current Market Description 273
Anahid Basiri, Terry Moore, Chris Hill and Paul Bhatia

Reviewers

The production of this book would not have been possible without the professional help of our scientific committee. We would like to thank all the following experts who have helped to review the papers published in this book.

Suchith Anand, University of Nottingham, UK
Gennady Andrienko, Fraunhofer IAIS/City University London, Germany/UK
Thierry Badard, Laval University, Canada
Rex Cammack, University of Nebraska at Omaha, USA
Cristina Capineri, Siena University, Italy
William Cartwright, RMIT University, Australia
Matt Duckham, University of Melbourne, Australia
Claire Ellul, University College London, UK
Peter Fröhlich, Telecommunications Research Center Vienna, Austria
Georg Gartner, Vienna University of Technology, Austria
Haosheng Huang, Vienna University of Technology, Austria
Mike Jackson, University of Nottingham, UK
Bin Jiang, University of Gävle, Sweden
Jie Jiang, National Geomatics Center of China, China
Markus Jobst, Austrian Federal Office for Metrology and Surveying, Austria
Hassan Karimi, University of Pittsburgh, USA
Farid Karimipour, University of Tehran, Iran
Jukka Krisp, University of Augsburg, Germany
John Krumm, Microsoft Research, USA
Chun Liu, Tongji University, China
Liqiu Meng, Technische Universität München, Germany
Xiaolin Meng, University of Nottingham, UK
Peter Mooney, National University of Ireland Maynooth, Ireland
Takashi Morita, Hosei University, Japan
Hans-Berndt Neuner, Vienna University of Technology, Austria
Ed Parsons, Google, UK
Michael Peterson, University of Nebraska at Omaha, USA

Martin Raubal, Swiss Federal Institute of Technology Zurich, Switzerland
Karl Rehrl, Salzburg Research, Austria
Günther Retscher, Vienna University of Technology, Austria
Tapani Sarjakoski, Finnish Geodetic Institute, Finland
Manuela Schmidt, Vienna University of Technology, Austria
Johannes Schöning, Hasselt University, Belgium
Volker Schwieger, University of Stuttgart, Germany
Stefan van der Spek, Delft University of Technology, The Netherlands
Josef Strobl, University of Salzburg, Austria
Nico van de Weghe, Ghent University, Belgium
Stephan Winter, University of Melbourne, Australia
Kefei Zhang, RMIT University, Australia
Sisi Zlatanova, Delft University of Technology, The Netherlands

Part I
Wayfinding and Navigation

Is OSM Good Enough for Vehicle Routing? A Study Comparing Street Networks in Vienna

Anita Graser, Markus Straub and Melitta Dragaschnig

Abstract As a result of OpenStreetMap's (OSM) openness and wide availability, there is increasing interest in using OSM street network data in routing applications. But due to the heterogeneous nature of Volunteered Geographic Information (VGI) in general and OSM in particular, there is no universally valid answer to questions about the quality of these data sources. In this paper we address the lack of systematic analyses of the quality of the OSM street network for vehicle routing and the effects of using OSM rather than proprietary street networks in vehicle routing applications. We propose a method to evaluate the quality of street networks for vehicle routing purposes which compares relevant street network features as well as computed route lengths and geometries using the Hausdorff distance. The results of our case study comparing OSM and the official Austrian reference graph in the city of Vienna show close agreement of one-way street and turn restriction information. Comparisons of 99,000 route pairs with an average length of 6,812 m show promising results for vehicle routing applications with OSM, especially for route length computation where we found median absolute length differences of 1.0 %.

Keywords OpenStreetMap (OSM) · Volunteered geographic information (VGI) · Quality assessment · Routing · Street networks

1 Introduction

Vehicle routing applications used in route planning, navigation, and fleet management software depend heavily on the quality of the underlying street network data. Errors such as missing streets, wrong or missing turn restrictions or one-way street information lead to wrong route choices and wrong distance estimations.

A. Graser (✉) · M. Straub · M. Dragaschnig
Dynamic Transportation Systems, Mobility Department, AIT Austrian Institute of Technology, Vienna, Austria
e-mail: anita.graser@ait.ac.at

G. Gartner and H. Huang (eds.), *Progress in Location-Based Services 2014*,
Lecture Notes in Geoinformation and Cartography,
DOI 10.1007/978-3-319-11879-6_1

Historically, street network data for vehicle routing applications was only available through a limited number of vendors or official government sources. With the increasing popularity of Volunteered Geographic Information (VGI) projects such as OSM, there is a growing interest in using such free and open data sources in routing applications for different modes of transport.

The adoption of OSM in professional settings is hindered by concerns about the unknown quality of OSM data. Besides simple omission of objects, potential users are also concerned about active vandalism. One important factor is that OSM quality is not consistent between regions. Some countries such as Germany and Austria have large communities of contributors (Neis 2012), which has been found to correlate positively with higher data quality (Neis et al. 2012), while other countries have only smaller groups of contributors. Additionally, OSM quality shows an urban-rural divide (Thaller 2009; Zielstra and Zipf 2010) with better quality in urban regions. It is therefore necessary to evaluate the map in the area of interest with respect to quality for a specific application before OSM can be used. Existing studies describe evaluation methods for quality aspects such as positional accuracy and attribute completeness (for an extensive, non-exhaustive list see OSM Wiki 2014). However, to date there has been no systematic analysis of the quality of the OSM street network for vehicle routing and the effects of switching vehicle routing applications from established proprietary or governmental street networks to OSM.

Aiming to fill this gap, this paper presents a comparative method which evaluates OSM street network quality in comparison to a reference street network. The method is based on comparisons of street network features such as turn restrictions and one-way streets as well as comparisons of routes calculated on both street networks. Route comparisons analyze both route length and route geometry to evaluate the effects of different street networks on the results of vehicle routing applications.

Section 2 presents an overview of related work, followed in Sect. 3 by a description of the methodology used in this study. Section 4 presents results of the comparison of OSM and the official Austrian reference graph in the analysis area. The paper closes in Sect. 5 with a discussion of results and an outlook for future work.

2 Related Work

A first systematic approach for analyzing the quality of OSM is presented by Haklay (2010). He compares OSM to Ordnance Survey UK calculating positional accuracy and comparing network length in regular grid cells covering the study area. Ather (2009) extends Haklay's work, comparing completeness of street names. Subsequent papers compare OSM to other street network datasets such as Ordnance Survey Ireland (Ciepluch et al. 2011), TeleAtlas/TomTom Multi-Net (Thaller 2009; Zielstra and Zipf 2010), or Navteq (Ludwig et al. 2011).

Ludwig et al. (2011) and Koukoletsos et al. (2012) follow a different approach based on matching street objects of OSM and street objects in a reference dataset. This enables them to compare attributes and geometries of specific street objects in two network datasets. Other studies, such as Neis et al. (2012), present OSM-internal checks on validity and topology of the street network.

Work focusing on routing-specific aspects of street networks includes Thaller (2009), who compared three routing examples calculated by OpenRouteService (which uses OSM) and TomTom personal navigation devices, based on route length, travel time estimation, and route choice.

Zielstra and Hochmair H (2012) compare shortest path lengths for 1,000 pedestrian routes of "typical pedestrian walking distance" per city. Routes were calculated using OSM and other free and commercial street network datasets (TomTom, Navteq, TIGER, ATKIS, as well as a combination of networks) in two German and two US cities. They found that OSM provided the most complete data source and the shortest routes—only outperformed by a combination of all available datasets.

On the topic of vehicle routing, Ludwig et al. (2011) found that the "oneway" attribute was missing from 28.1 % of features in inhabited areas and 48.8 % of features in uninhabited areas of Germany when compared to the Navteq street work. Similarly, "speed limit" was found missing for 80.7 % of objects in inhabited areas and 92.6 % of objects in uninhabited areas. Neis et al. (2012) compared the number of turn restrictions found in OSM and TomTom and showed that the number of turn restrictions in OSM is significantly lower (21,000 instead of 176,000 for Germany in June 2011) than in TomTom MultiNet dataset.

As shown in Graser et al. (2014), comparing the number of turn restrictions can cause misleading results due to differences in how street networks are modeled with respect to network generalization and representation of driving restrictions. While a comparison of turn restriction counts between the official Austrian reference graph "Graph Integration Platform" (GIP) and OSM for the greater Vienna region in December 2012 found 2,500 turn restrictions in GIP but only 691 (27.6 %) in OSM, a systematic routing-based comparison showed that 1,515 (60.6 %) of the 2,500 GIP turn restrictions had a matching representation in OSM. Similarly, 10,499 (87.8 %) of the 11,964 one-way streets in GIP could be matched to one-way streets in OSM.

None of the studies so far offer a systematic evaluation of how exchanging an established street network dataset with OSM affects the output of vehicle routing applications with respect to resulting route length and route geometry.

3 Methodology

Our approach comprises the following steps: After the initial preparation of routable graphs, we compare the street networks based on network completeness, similarity of turn restriction and one-way street information, and vehicle routing results.

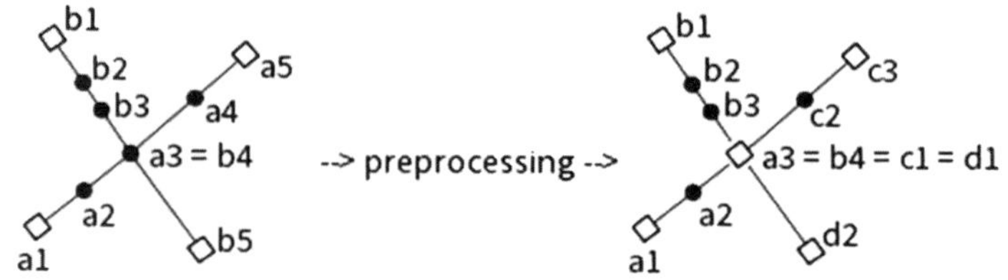

Fig. 1 OSM preprocessing step splitting edges at intersections

3.1 Preparing Routable Graphs

The OSM street network consists of nodes and edges but OSM edges are not necessarily split at each intersection. Instead, in the OSM representation, edges are considered to be connected if they share a common node at the point of intersection. Therefore, OSM is not routable without preprocessing, which splits edges at the appropriate intersections as depicted in Fig. 1. The GIP street network, on the other hand, is modeled using nodes and edges which are split at intersections and connected through explicit turn relations. Without a turn relation, even GIP edges sharing a common intersection node are not considered to be connected. For more details on the different approaches to street network modeling used in OSM and GIP, including a matching of OSM highway tag values and GIP functional road classes, see Graser et al. (2014).

Another aspect where the modeling approaches of OSM and GIP differ is the handling of features such as driving permissions and turn relations. While GIP tends to explicitly define driving permissions for various modes of transport, OSM tends to use conventions and explicit restrictions; i.e. the OSM tag combination vehicle = no and bicycle = yes evaluates to a ban on all vehicles except bicycles. Similarly, GIP turn restrictions are modeled implicitly through missing turn relations, while in OSM, all turn maneuvers are allowed at an intersection as long as there is no explicit restriction relation specified.

In order to create a correct OSM routing graph, edges that share a node at their intersection have to be split up at the intersection node. It is worth noting that the data preparation has to ensure that edges which do not share a common node at the point of intersection are not split in order to avoid creating junctions where there should not be any. This is especially relevant at overpasses and underpasses created by bridges, tunnels or similar network features. Turn restrictions are created from OSM tag combinations that define turn maneuvers and finally, each driving direction and turn restriction is labeled with the modes of transport they concern.

3.2 Street Network Comparison

The comparison of street networks is divided into three parts: (1) assessment of street network completeness; (2) comparison of turn restriction and one-way street information relevant for vehicle routing; (3) comparison of routing results.

The first step is a general comparison of the length of the street networks of OSM and GIP determined by calculating the total sum of the length values of all street network graph edges. This is the most common test for **completeness of street networks** used in Haklay (2010) and numerous subsequent publications. This test can only provide a rough estimate of data completeness since it assumes that both datasets contain similar types of information. Before applying the test, it is therefore necessary to remove road classes which are not represented in both datasets.

In a second step, we compare **turn restriction and one-way street** information relevant for vehicle routing applications. The routing-based comparison method presented in Graser et al. (2014) compares forbidden maneuvers of driving against the one-way street direction and turning at a turn restriction of one street network with routing results calculated on the second street network (see Fig. 2) to test whether both street networks contain matching driving restrictions. Similarity between forbidden turn maneuver and route generated on the second street network is determined using the Hausdorff distance (Hausdorff 1914). A Hausdorff distance above 10 m is interpreted as a correctly modeled turn restriction. Additionally, similarity between a maneuver describing driving against the one-way direction and the route generated on the second street network is determined using length comparison. If the one-way information is present in both street networks, the generated route has to find a way around the driving restriction and will therefore be considerably longer than the forbidden maneuver, which is generated by extracting a 10 m long section from the center of a one-way street (see forbidden maneuver in Fig. 2c, d). A route length above 20 m is interpreted as a correctly modeled one-way street.

The **routing comparison** step of the street network comparison procedure examines routes calculated between identical start and end points. A regular grid is created and used to distribute start and end points in the study area. For each cell

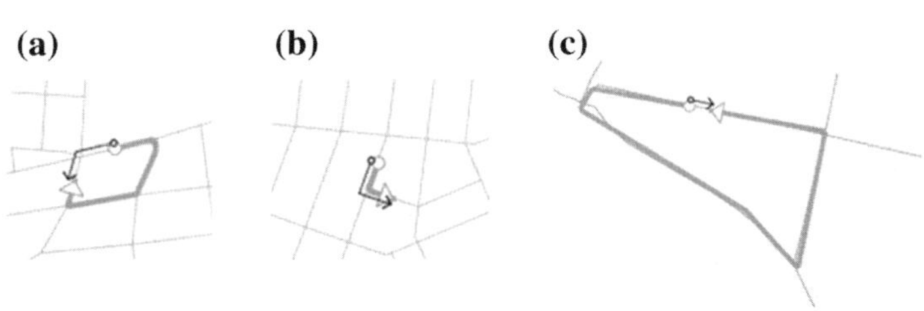

Fig. 2 **a** Correctly modeled turn restriction; **b** Missing turn restriction; **c** Correctly modeled one-way street; **d** Missing one-way restrictions; (*narrow black arrows*: forbidden maneuver; *wide grey arrows*: routes generated on the comparison graph)

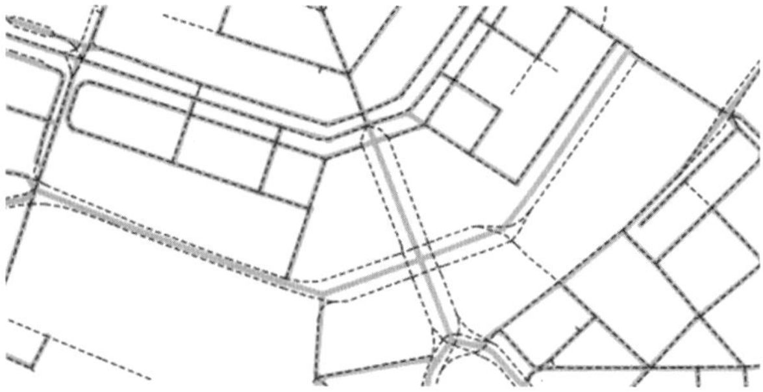

Fig. 3 Details and network generalization differences between OSM (*dashed black lines*) and GIP (*wide grey lines*) at Schwarzenbergplatz

pair consisting of a source and target cell, *n* routes are calculated. Before the routes are calculated, it is necessary to select route start and end points. Distributing start or end points randomly within the cells would lead to ambiguous situations, e.g. if the points end up in the middle between two edges or at an intersection where it is unclear which street should be selected for the route start or end. To minimize these ambiguous situations, we first choose a random network edge within the cell and then select the center point of the edge as start or end point for the route. To select start and end points in the second dataset, a simple map matching routine is applied as follows: the start and end points generated on the first dataset are each matched onto the 13 nearest junctions in the graph and of all incoming and outgoing edges of these junctions, the one with the minimum normal distance is chosen. Finally, the routes are calculated using shortest distance routing with Dijkstra's algorithm (Dijkstra 1959).

The evaluation starts by computing length differences between the routing results on OSM and the reference graph. The distribution of length differences provides a first assessment (see also Fig. 5b). Systematic differences can be observed if routes in one network are systematically shorter than in the other network. Systematic differences might be due to (1) higher road density in one network; (2) lack of driving restrictions; or (3) lack of necessary connections.

Even if the resulting length differences are small or non-existent, this evaluation step only confirms or disproves that route calculations on both graphs result in routes of the same length. While this might be sufficient for certain kinds of vehicle routing applications which only focus on the resulting distance estimates, further evaluations are necessary for applications which depend on calculating correct route geometries, because the first evaluation step cannot confirm whether both route calculations result in the same routes in respect to route geometry. Therefore, in the second part of the routing comparison procedure, the Hausdorff distance is calculated to assess route similarity based on route geometry since it describes the difference between two route geometries independent of the route length.

4 Results

The datasets used in this study are the raw OSM XML data provided by Geofabrik (2013) for March, 19th 2013 and the effective GIP export for routing motorized traffic (called “MIV export”) within a 10 × 10 km study area (Fig. 4). In the following comparisons, the GIP export is used as the reference street network graph which OSM is compared with. We want to point out that the methodology could just as well be applied to commercial street network data by providers such as TomTom or Nokia HERE. Since the GIP export does not contain unpaved roads which would be equivalent to the OSM type “track”, we removed streets of type “track” from the OSM network.

4.1 Street Network Completeness

A preliminary assessment of OSM and GIP street networks shows that the OSM street network is 1,402 km long and thus 210 km (+17.6 %) longer than the GIP export which is 1,192 km long. Since by removing unpaved roads we ensured that both datasets represent the same road classes, this difference is largely due to the more generalized nature of the GIP export street network which is optimized for routing applications and only contains road center lines as shown in Fig. 3. No generalization was applied to the OSM dataset.

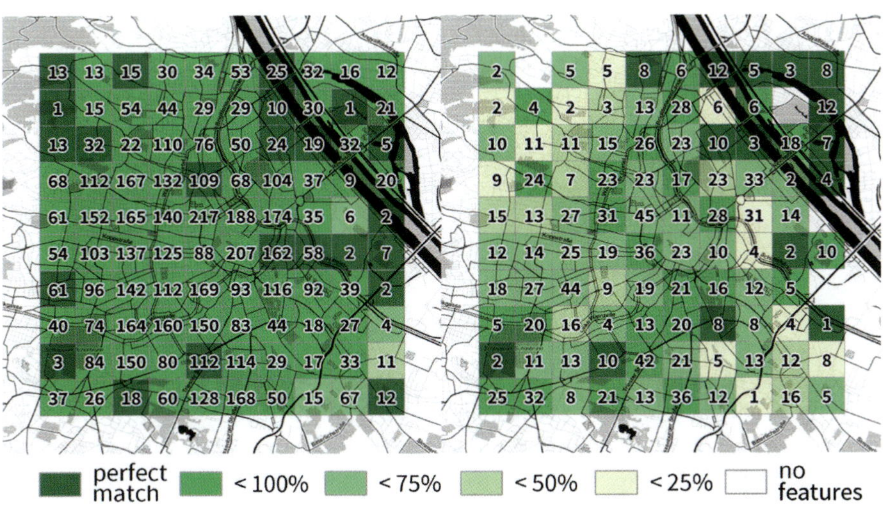

Fig. 4 Spatial distribution of matching one-way streets (*left*), and turn restrictions (*right*); absolute counts (value in the cell) and ratio of matching features (*color*)

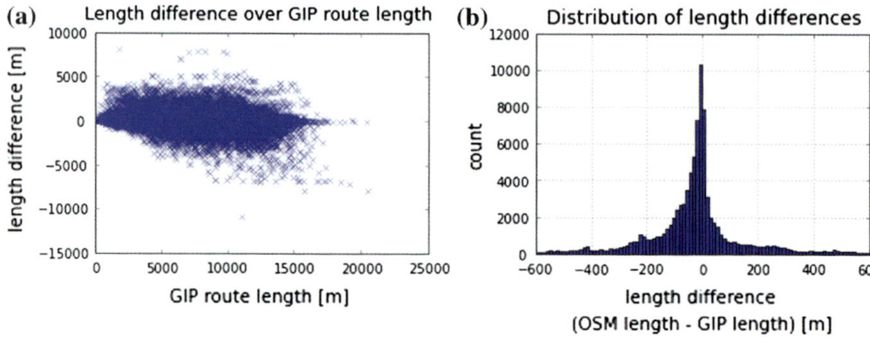

Fig. 5 **a** Length differences depending on GIP route length for individual routes; **b** Distribution of length differences for individual routes

4.2 One-Way Streets and Turn Restrictions

The comparison of one-way streets shows that 6,289 (95.4 %) of the 6,595 GIP one-way streets in the study area can be matched to a one-way street in OSM. Similarly, 842 (68.3 %) of the 1,232 GIP turn restrictions have a matching representation in OSM.

Figure 4 depicts the spatial distribution of one-way streets and turn restrictions in the analysis area. The rate of matching features is color-coded using darker shades for cells with more matches and lighter shades for cells with fewer. The number written inside the cell states the number of occurring one-way streets or turn restrictions in the respective cell. The numbers in the turn restriction map clearly show that turn restrictions are much less common than one-way streets. Additionally, some cells do not contain a single one and are therefore omitted from the turn restriction map. While agreement about one-way streets is high with 91 out of 100 cells with a match ratio better than 80 %, only 30 of the 96 cells which contain turn restrictions reach the same ratio of 80 % matching features in OSM and GIP.

4.3 Routing Comparison

In this case study, we used a grid with 100 1 × 1 km cells and calculated ten route pairs consisting of an OSM route and a GIP route per cell pair. This leads to a total of 99,000 route pairs with an average GIP route length of 6,812 m (min: 54 m; max: 20,465 m). We calculate each route pair's length difference as OSM route length minus GIP route length. Negative difference values therefore stand for shorter OSM routes. Based on all 99,000 routes, the mean length difference is −15.5 m and the median length difference –17.3 m. These results show that OSM routes tend to be shorter than the corresponding GIP routes. Figure 5a depicts the relation of length

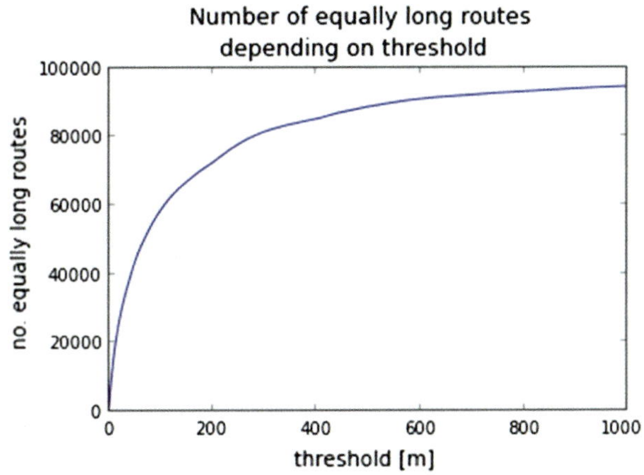

Fig. 6 Number of route pairs with equally long OSM and GIP routes depending on threshold

difference and GIP route length per route pair. High negative length differences are found for long GIP routes over 5 km length while positive length differences are also found for shorter routes. Figure 5b depicts the overall distribution of length differences per route pair and clearly shows the trend of shorter OSM routes in the shift towards negative length differences.

We have seen the trend towards shorter OSM routes, but how often can OSM and GIP routes be considered equally long for the purpose of vehicle routing applications? Figure 6 presents the number of route pairs with equally long OSM and GIP routes depending on the threshold chosen to define "equally long": at a threshold tolerance of ±10 m, 15,874 (16.0 %) of the total 99,000 routes are considered equally long. For a threshold of ±25 m this value rises to 29,325 (29.6 %), growing to 58,373 (59.0 %) for a threshold of ±100 m. Additional evaluations of absolute length differences in relation to GIP route length show that the median OSM route length deviates by 1.0 % from the corresponding GIP route length.

To gain a better understanding of the spatial distribution of route pairs with similar route length and those with bigger deviations, we further evaluate the length difference values grouped by route starting cell. Mean route length by cell varies between 5 and 9 km depending on whether the cell is located in the center of the analysis area or around its border. Figure 7b shows that, in most cells, OSM routes are shorter than GIP routes, confirming our previous interpretation of individual route results, while Fig. 7a depicts the same median length difference values plotted over the median GIP route length. This confirms the intuition that higher length difference values are found for cells with longer GIP routes.

Figure 8 depicts the length differences for all route pairs starting in the respective cell. Cells in the center of the grid generally show lower median length difference values than cells around the border of the analysis area. While cells with high negative median difference values—indicating that OSM routes starting there are considerably shorter than the respective GIP routes-cluster in the northwest, cells

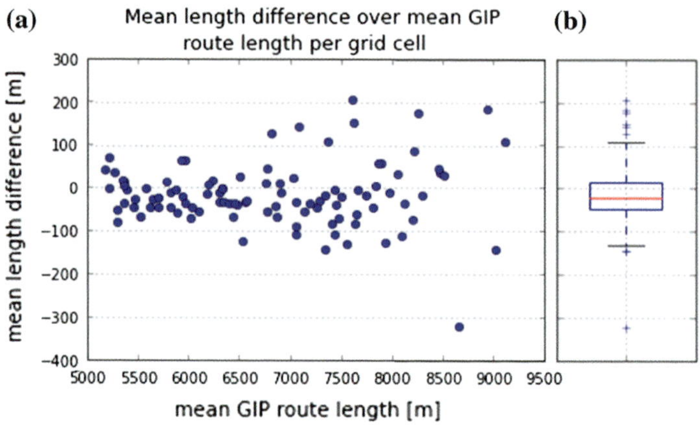

Fig. 7 **a** Median length difference over median GIP route length per route starting grid cell; **b** Distribution of median length differences per route starting grid cell

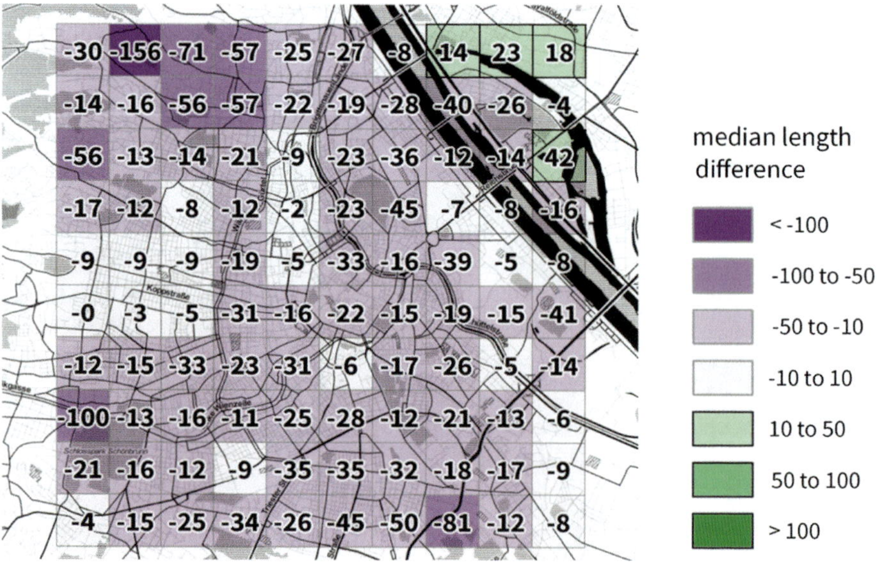

Fig. 8 Median length difference in meters for all route pairs starting in the given cell

with positive values are found in the northeast of the analysis area, north of the river Danube where most of the routes have to cross the Danube bridges to the southern part of Vienna. In any case, it has to be noted that route length differences are accumulated along the whole route, and the underlying street network deviations causing the differences are therefore not necessarily located in the route starting cell.

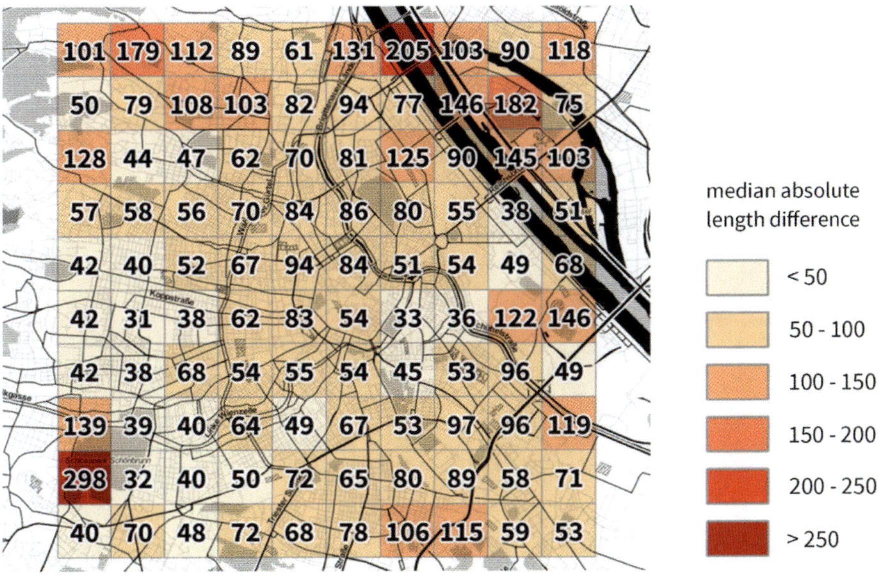

Fig. 9 Median absolute length difference in meters for all route pairs starting in the given cell

Figure 9 shows the median absolute length difference thus highlighting areas with bigger length differences independent of whether OSM or GIP routes tend to be shorter. As before, higher values are found at the borders of the analysis area but the focus of highest difference values has shifted to areas southwest of Schlosspark Schönbrunn (298 m difference) and along the Danube (205 m difference). Closer inspection of the routes starting in the grid cell near this park shows that the cell is mostly covered by the park and contains only a very limited number of street edges (< 20 in either datasets). As a result, the algorithm picking route start and end points ends up picking from this small set of edges over and over for each cell pair and thus a single network difference can affect multiple routes.

To gain a better understanding of the sources of the length differences, 25 routes were inspected manually. For all randomly selected and inspected routes with very large differences (eight kilometers) the reasons were that the automatic matching process selected topologically different start or end edges, e.g. motorway links instead of motorway exit. For further studies, it is recommended to take special care that the chosen start and end edges of the routes match to the same logical edges in the road graphs.

Other causes for length differences were map defects or inaccuracies, such as missing or wrong information about where motor vehicles are allowed to drive, different one-way information, missing or wrong turn restrictions, and different lengths of dead end streets.

After these length-based comparisons, the following sections present the results of comparing route geometries. In this study, similarity of route pair geometries was

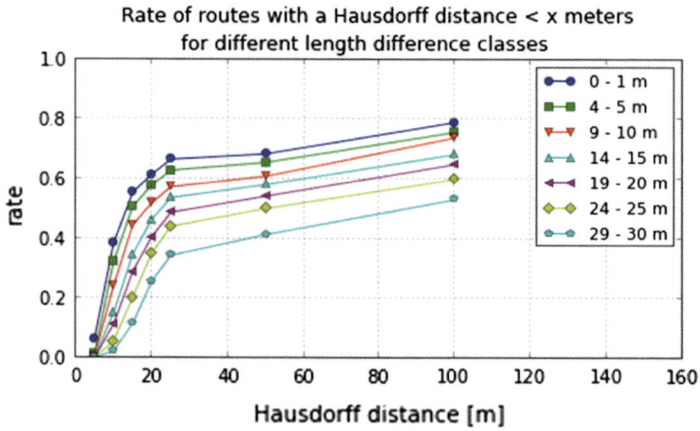

Fig. 10 Rate of routes with a Hausdorff distance under a given threshold

calculated as the Hausdorff distance between OSM and GIP route since it describes the difference between two route geometries independent of the route length. Figure 10 shows how the rate of route pairs with OSM and GIP route geometries which are considered similar grows as the Hausdorff distance threshold, which is used to define "similar", is increased. The figure shows separate curves for several length difference classes from the top-most curve which represents only route pairs with a length difference of 0–1 m to the curve at the bottom which represents route pairs with a length difference of 29–30 m. These results show a steady increase of similar routes up to a threshold of 25 m which slows down considerably afterwards.

Based on these results, Fig. 11 depicts the relation between length difference of OSM and GIP routes and the rate of route pairs with similar route geometries defined by a Hausdorff distance less than 25 m. A threshold of 25 m was chosen since it is well below the size of typical city blocks, thus eliminating routes which deviate by one city block while allowing for smaller deviations caused by different levels of street network generalization. The graph in Fig. 11 shows a linear

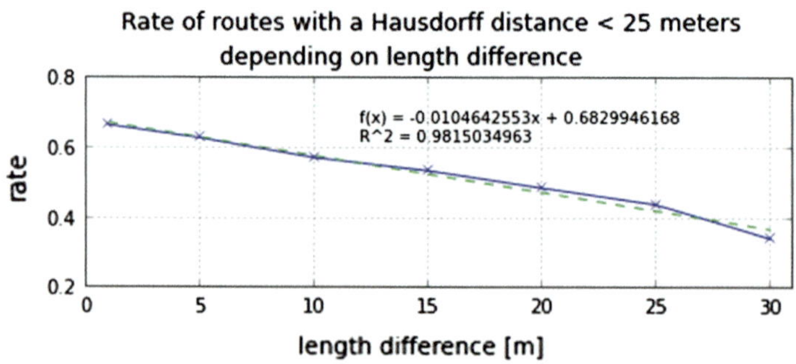

Fig. 11 Rate of routes with a Hausdorff distance under 25 m over length difference

relationship with a high coefficient of correlation $R^2 = 0.98$. Since calculating Hausdorff distance is computationally expensive, we use the resulting linear relationship

$$-0.0104642553 * \text{absolute\_length\_difference} + 0.6829946168 \quad (1)$$

to estimate the total number of route pairs with both a length difference less than 25 m and a Hausdorff distance less than 25 m. Of the total 99,000 routes in the sample, 16,903 (17.1 %) route pairs fall into this class and are therefore considered to be a perfect match for the purpose of this study.

5 Conclusion and Future Work

The comparison of OSM and GIP for vehicle routing application presented in this work investigates the influence of switching from GIP to OSM on resulting shortest path route length and route geometry. The study covers comparisons of street network completeness, turn restriction and one-way street information as well as 99,000 routes with a mean GIP route length of 6,800 m in the city of Vienna.

Route length comparison results show that for 59.0 % of routes, the computed OSM route length is within a tolerance of 100 m of the corresponding GIP route length, and for 29.6 % of routes, OSM route length is within a tolerance of 25 m of the corresponding GIP route length. Observed route length differences vary by location but it has to be noted that route length differences are accumulated along the whole route and, as a result, locating the street network deviations causing the differences therefore is no trivial task. OSM routes tend to be shorter than GIP routes which could be explained by two factors: first, the OSM network could be denser than the GIP network and thus contain more "shortcuts", be they right or wrong; and second, the OSM street network could contain fewer driving restrictions and thus be more connected. While comparisons of street network length show that the OSM street network within the analysis area is 17.6 % longer than the GIP street network, the differences are mostly due to the more generalized nature of the GIP export and not due to additional connecting streets in the OSM street network. Regarding driving restrictions, a comparison of one-way streets shows that 95.4 % of the 6,595 GIP one-way streets in the study area can be matched to a one-way street in OSM and similarly, 68.3 % of the 1,232 GIP turn restrictions have a matching representation in OSM. Differences in the remaining turn restrictions and one-way streets will influence route length and geometry deviations. Based on the results of an evaluation of absolute length differences relative to GIP route length, vehicle routing applications that compute route length based on OSM instead of GIP would result in routes with a median absolute length difference of 1.0 % relative to the original GIP route length.

To further evaluate the similarity of routing results, Hausdorff distance between OSM route geometry and the corresponding GIP route geometry was calculated.

Results of this evaluation show that 17.1 % of route pairs have both a length difference less than 25 m as well as a Hausdorff distance less than 25 m and are therefore considered to be a perfect match for the purpose of this study. It has to be noted that due to the varying quality of OSM, applying the analysis methods presented in this study in other geographic regions might result in significantly different results.

Expected correlations between shorter route length and a better agreement of routing results both in respect to route length and geometry should be investigated in subsequent studies. Further work is planned to evaluate the effect of migrating to OSM on specific vehicle routing applications, such as floating car data systems that require routing between successive vehicle positions, which are sampled at intervals up to two minutes, leading to considerably shorter routes than the ones evaluated in this study.

Acknowledgements This work is funded by the Austrian Climate and Energy Fund within the "Electric Mobility flagship Projects" program (project "emporA 2") and the KLIEN initiative of the Austrian bmvit (project no. 839,478 "Crossing Borders").

References

Ather A (2009) A Quality Analysis of OpenStreetMap Data. Master's thesis, University College London

Ciepluch B, Mooney P, Jacob R, Zheng J, Winstanely AC (2011) Assessing the quality of open spatial data for mobile location-based services research and applications. Arch Photogrammetry, Cartography Remote Sens 22:105–116

Dijkstra EW (1959) A note on two problems in connexion with graphs. Numer Math 1:269–271

Geofabrik (2013) http://www.geofabrik.de/data/download.html. Accessed 19 March 2013

Graser A, Straub M, Dragaschnig M (2014) Towards an open source analysis toolbox for street network comparison: indicators, tools and results of a comparison of OSM and the official Austrian reference graph. Transactions in GIS 18:510–526. doi:10.1111/tgis.12061

Haklay M (2010) How good is volunteered geographical information? A comparative study of OpenStreetMap and Ordnance Survey datasets. Environ Plan 37(4):682–703. doi:10.1068/b35097

Hausdorff F (1914) Grundzüge der Mengenlehre

Koukoletsos T, Haklay M, Ellul C (2012) Assessing Data Completeness of VGI through an Automated Matching Procedure for Linear Data. Trans GIS 16(4):477–498

Ludwig I, Voss A, Krause-Traudes M (2011) A Comparison of the Street Networks of Navteq and OSM in Germany. Springer, Advancing Geoinformation Science for a Changing World, pp 65–84

Neis P (2012) Distribution of Active Users in OpenStreetMap – Oct-Nov 2012. http://neis-one.org/2012/11/active-users-osm-nov12/. Accessed 11 May 2014

Neis P, Zielstra D, Zipf A (2012) The Street Network Evolution of Crowsourced Maps: OpenStreetMap in Germany 2007-2011. Future Internet 4:1–21. doi:10.3390/fi4010001

Thaller D (2009) Die Open-Source-Plattform "OpenStreetMap", eine Konkurrenz für Geodatenhersteller?. Master's thesis, Universität Wien. Fakultät für Geowissenschaften, Geographie und Astronomie

OSM Wiki (2014) Research. http://wiki.openstreetmap.org/w/index.php?title=Research&oldid=995511. Accessed 10 May 2014

Zielstra D, Hochmair H H (2012) Comparison of Shortest Path Lengths for Pedestrian Routing in Street Networks Using Free and Proprietary Data. Proceedings of Transportation Research Board–Annual Meeting, Washington, DC, USA, 22–26 January 2012

Zielstra D, Zipf A (2010) A Comparative Study of Proprietary Geodata and Volunteered Geographic Information for Germany. AGILE 2010. Guimaraes, Portugal

Calculating Route Probability from Uncertain Origins to a Destination

Carolin von Groote-Bidlingmaier, David Jonietz and Sabine Timpf

Abstract Uncertainty in location information can affect the results of network-based route calculations to a high degree. In this study, a routing scenario is analysed where the destination is known, but the location of the point of origin can only approximately be described as "somewhere inside a polygon". Using the concrete example of car-driving football fans arriving at a game, an approach is proposed to compute and describe the probability of them taking specific routes from their home county to the stadium. A set of candidate points of origin is created and shortest paths to the destination calculated. The observed frequency of an edge being included in a route allows inferring a routing probability for each edge. Several methods to derive a set of candidate points of origin are presented and discussed, ranging from purely geometrical to geographically weighted approaches. Our results show that the differences between the methods in determining the points of origin produce only slightly different probabilities, i.e., neither advantages nor drawbacks are to be expected from using a purely geometrical approach.

Keywords Routing · Probability · Network analysis

1 Introduction

The fact that almost all spatial information is marked by some degree of uncertainty poses particular challenges for its further use, for instance in the context of spatial analysis. The sources of uncertainty are manifold, ranging from data collection to

C. von Groote-Bidlingmaier (✉) · D. Jonietz · S. Timpf
Geoinformatics Group, Department of Geography, University of Augsburg, Augsburg, Germany
e-mail: carolin.vongroote-bidlingmaier@geo.uni-augsburg.de

D. Jonietz
e-mail: david.jonietz@geo.uni-augsburg.de

S. Timpf
e-mail: sabine.timpf@geo.uni-augsburg.de

G. Gartner and H. Huang (eds.), *Progress in Location-Based Services 2014*,
Lecture Notes in Geoinformation and Cartography,
DOI 10.1007/978-3-319-11879-6_2

analytical methods. One can distinguish between different types of uncertainty, including inaccuracy and error, vagueness and incompleteness (Worboys 1998). In the context of location-based services (LBS), for instance, positioning errors could result from poor GPS accuracy, conceptual vagueness from the use of imprecise spatial statements such as "near" or "far" or incompleteness in the case of missing spatial information, such as network attributes in a routing task (Basiri et al. 2012).

1.1 Uncertain Routing

Routing is included in the functionality of many LBS as well as a basic task for transportation planners and involves the use of algorithms to identify an optimized path from an origin to a destination through a weighted network. While in the context of LBS, the focus is typically on the individual traveller, transportation modelling generally aims at forecasting travel demand and its resulting effects such as traffic volumes at a more aggregate level (MacNally 2007). In both cases, depending on the optimization criteria, possible solutions can include the shortest, fastest, cheapest or other paths (Miller and Shaw 2001).

With regards to uncertain data, research has so far focused on edge attributes (e.g., Liao et al. 2014), destination points (e.g., Qiu et al. 2013) as well as positional uncertainty (e.g., Gonzales and Stenz 2007; Hait et al. 1999). The paper is situated in this thematic context but starts from a different flavour of uncertainty. A routing scenario is discussed which focuses on identifying the most probable route from an area of origin to a predefined destination. While the destination is known and can be represented as a fixed node on the street network, the origin of the route can only be approximately described as a distinct polygon due to either incomplete or inaccurate spatial information. In fact, a very similar challenge is posed to transportation planners during the process of route assignment, when trips are modelled for distinct traffic analysis zones (TAZ) and assigned to the road network in order to predict traffic volumes (MacNally 2007). In this case, points of origin are seldom known (e.g., home locations), but in most cases approximated as the centroid of the TAZ is used for the routing (Qian and Zhang 2012). In the past, there has also been work on the possible influences of using a population- or household-weighted centroid instead of a purely geometric one (Chang et al. 2002). In the case study presented in this paper, the aim is to determine the most probable route taken by car-driving visitors to a large-scale event, in this case a football game at a stadium. The exact origins of the visitors are unknown, although information about their home county can be retrieved from their license plates. The problem of calculating probable routes from an uncertain origin, however, is not just restricted to transportation planning, but could also be applied to positional inaccuracy in the context of LBS, i.e., using the error ellipse (circle) of uncertain GPS positioning.

In this paper we propose the use of a large set of candidate points instead of merely one representative centroid. We introduce a geographically weighted approach, which we compare with purely geometric methods in order to evaluate its appropriateness and practicality for predicting the usage of specific routes to a stadium for a mass event. In contrast to determining a most probable starting point and calculating a single route, as in the case of TAZ centroids, our approach has the advantage to produce probabilities for each segment of all routes leading to the destination. Thus, without the need for detailed data about exact home locations, planners and mass event managers may identify potential bottlenecks and reduce congestion by applying appropriate measures and LBS may take this knowledge into account when suggesting a specific route.

1.2 Approaches to Uncertainty in Spatial Data

Today, the term uncertainty is increasingly used instead of error to describe the difference between spatial data and their corresponding real world entities, thereby acknowledging the fact that representations are always merely approximations of the truth (Zhang and Goodchild 2002). Except for geographical abstraction, uncertainty may also result from processes of approximation, measurement or generalization (Goodchild 2009). At the same time, however, users of GIS are often unaware of these issues, a fact which may at worst lead to decisions being made on the basis of questionable spatial information. As a result, extensive research has focused on measuring, modelling, visualizing and analysing the propagation of uncertainty in spatial data.

In the specific context of modelling uncertainty, a range of theoretical frameworks have been applied, among others probability theory and statistics, fuzzy set theory, rough sets, as well as possibility theory. While fuzzy and rough sets are useful tools to extend set membership and set boundaries beyond crisp values, and possibility theory is concerned with measuring the feasibility of some event, probability theory describes methods to calculate its probability (Wang et al. 2005). Traditionally, values ranging from 0 (null event) to 1 (entire sample space Ω, i.e., every possible outcome) are used to denote the probability $p(E)$ of a particular event E (Jaynes 2003). Among several others, there is a most common view of $p(E)$, which is prevalent in statistics and relates it to the relative frequency $f(E)$ of the occurrence of E in a number of trials of an experiment, such that $p(E)$ equals $f(E)$ (Hajek 2012).

This paper is structured as follows: we first present the case study that inspired us to start this investigation. In Sect. 3, the method for calculating route frequency is described in detail showing the two steps of first determining potential points of origin and second calculating the probability from the frequency. Sect. 4 compares and discusses the results from the different methods and Sect. 5 presents conclusions and future work.

2 Case Study: Predicting Usage of Access Routes to a Stadium

The specific problem presented in this study is set in the context of traffic management for large-scale events. In our specific study area, authorities have to cope with the dense flow of visitors moving to and from the specific site, in our case a soccer stadium. Figure 1 shows the street network of the counties (Landkreise) Augsburg, Aichach–Friedberg, and Dachau as well as the location of the SGL arena in the south of Augsburg.

Despite increased provision and marketing of public transport, many visitors still arrive in their own car and within a relatively short time interval. Thus, volumes of motorized traffic are multiple times higher than usual. In order to avoid negative effects such as traffic jams and prolonged waiting times at the main access points or unannounced shortages of parking space, it would be useful for planners to predict the visitors' access route to some degree, for instance in order to use traffic guidance systems to increase the efficiency of traffic flows in the vicinity of the stadium. For this, however, information about the visitors' access route is critical. Personal interviews are not practical, mainly due to the sheer mass of visitors. However, limited information about their home county can be retrieved from their license plates. In this case there are no origin-destination pairs, but the location of the origin

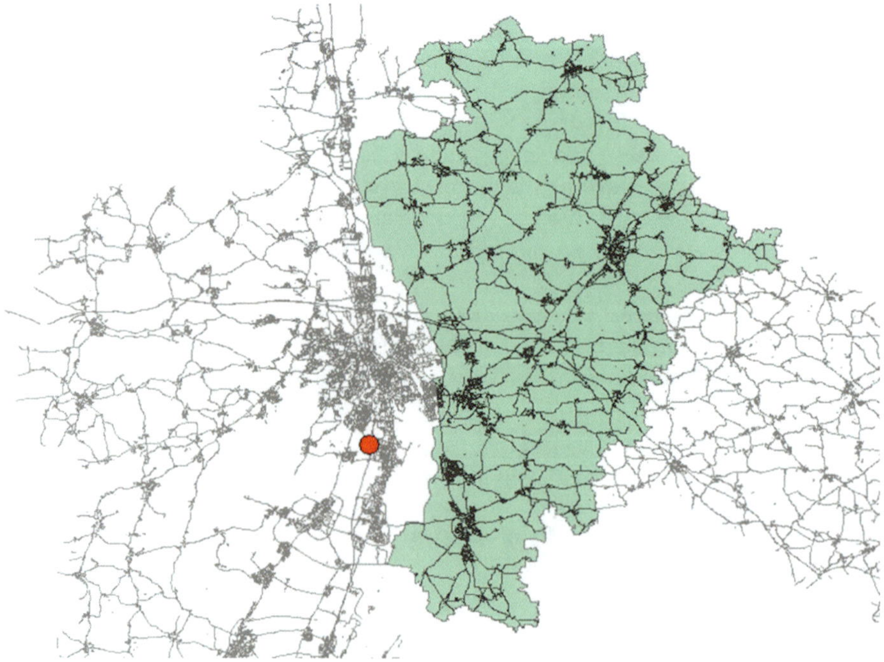

Fig. 1 Overview over the study area in Bavaria. The *red circle* indicates the SGL arena and the *green area* is the extent of the county Aichach–Friedberg, Bavaria

can be approximated using the administrative border of the respective county. Thus, predicting access routes poses a practical problem to the analyst. With regards to the mere extent and location of the polygon relative to the destination point, it can be argued that the use of a centroid as representative point of origin seems inappropriate in this case. In fact, our aim is to analyse the influence of different spatial distributions of the origins within the polygon of origin on the most probable route. In the following calculation we work with the assumption that the origin v' is supposed to be somewhere in Aichach–Friedberg and the destination v_d is the SGL arena in the south of Augsburg (red circle in Fig. 1).

3 Calculating Most Probable Routes

Based on the intention to identify the most probable route from an uncertain point of origin to a predefined destination, our proposed approach is as follows: We assume a street network represented in the form of a graph $G = (V, E)$. There is an unknown point of origin o ∈ V, the position of which is approximated by a set of points within a polygonal boundary P such that o ∈ P. Furthermore, there exists a known destination vertex $\mathrm{v_d}$, which is part of the set of points on the graph but outside the set of points representing the polygon: $\mathrm{v_d} \in \mathrm{V}$ and $\mathrm{v}_d \notin \mathrm{P}$. Since o is unknown, we introduce a subset $\mathrm{V}' \in \mathrm{V}$ and $V' \in P$ of vertices as candidate origins on the street network or projected onto the network and within the polygon, and compute a set of shortest paths $SP = \{v_0' e_1 v_1 e_2 \ldots e_k v_d \text{ where } e_i = v_{i-1}^{(\prime)} v_i, \forall 1 \le i \le k\}$ from each $v' \in \mathrm{V}'$ to d. Of course, the restriction to the network distance as single impedance represents a simplification, especially since a wide range of additional path optimization criteria such as time or easiness are used by humans during route planning. In addition, assuming that humans have perfect knowledge about the road network is certainly not fully realistic (Prato 2009). Considering the issue of model simplicity, however, as well as the fact that distance is in fact widely used as impedance in transportation models and thinking of the decision maker as a LBS, these limitations seem acceptable. On this basis, we calculate normalized values ranging from 0 to 1 for each $e \in E$ and e ∈ P to denote the relative frequency $f(e)$ of $e \in SP$. Following the most prevalent interpretation of probability, we argue that the higher the value of $f(e)$, the higher the probability $p(e)$ of e to be visited when a traveller starts a trip to v_d from an unknown o. The values $f(e_{1\ldots n})$ are used in the second step as impedance for a route calculation from s to the furthest v', in order to determine the most probable route for all member vertices of V'.

3.1 Methods of Generating Points of Origin V'

The results of the computation described above can be expected to depend to a high degree on certain characteristics of the vertex subset V', such as the number of vertices included, their mutual distance and their distribution on the road network. Moreover, in reality, the chance of being the point of origin for a trip is not equal among all v'. Rather, vertices located in areas of high population or road density can be assumed to be of a higher relative importance for the calculation than others, a fact which can be expressed in the form of individual weight coefficients assigned to specific vertices. In this study, a range of possible approaches is analysed, including purely geometrical and geographically weighted methods. For the purpose of this paper we chose 50 and 1,000 randomly distributed points on the network, regularly dispersed points at a distance of 500 and 1,000 m in a weighted and unweighted version (see Sect. 3.1.2), and points derived from street network density (see Sect. 3.1.3).

Independent from the actual method of creating V', the aim was to minimize the number of points in order to reduce the computational effort required for the analysis. Since the only impedance value used for the routing process will be network distance, it can be assumed that shortest paths will be similar for all v' which are located on the same edge e_i. Accordingly, if for a particular e_i the number of newly created vertices exceeds 1, i.e., $N(v') > 1$, they will be combined to one representative vertex and their absolute number stored as an attribute. In case of weight coefficients being assigned to v', the mean weight value will be calculated and saved as well, to make sure that the following calculations are not affected by any loss of information.

3.1.1 Randomly Distributed Points of Origin

As stated previously, among other potentially influential factors, we expect the results of the route calculations to depend on the number and distribution of v' as well. As a first approach, therefore, 50 and 1,000 random vertices were created on the network (see Fig. 2a, b), and used as input origins for a shortest path calculation to the destination v_d. To acknowledge the potential effect of the number of v', several trial runs were conducted.

According to the law of large numbers, a general principle which describes how, in a large number of trials, the observed frequency of an event will tend towards its theoretical probability, it can be expected that above a certain number of v', there will be no significant change in route frequency (Hazewinkel 2001). In this routing scenario, no additional weighting function for generating the vertices was applied. There is, however, an implicit influence of street network density, since on denser segments of the graph the chance of a point of origin being created is higher. The influence of population density will be included in the second approach, described in the following section.

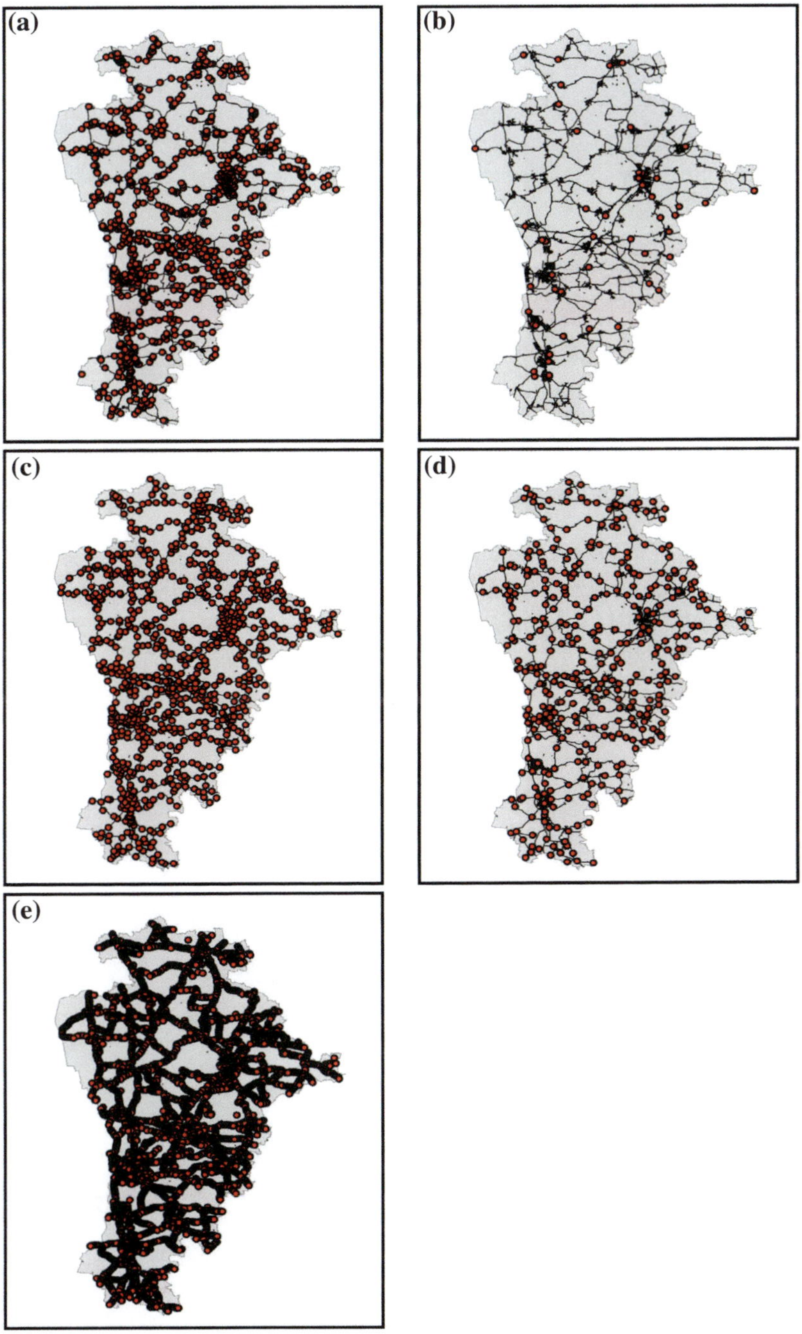

Fig. 2 Points of origin derived through different methods **a–e**, **a** randomly distributed points of origin-1,000 points, **b** randomly distributed points of origin-50 points, **c** regularly dispersed points of origin with 500 m, **d** regularly dispersed points of origin with 1,000 m, **e** points of origin derived from street network density

3.1.2 Regularly Dispersed Points of Origin

In contrast to a random distribution, for this approach an approximately even dispersion of points of origin was chosen. Vertices v' were created with a mutual distances of 500 and 1,000 meters on the graph edges $e \in \mathrm{E}$ and $e \in \mathrm{P}$ (see Fig. 2c, d). In order to avoid higher point densities in the vicinity of edge junctions, additional processing steps were necessary to arrive at a point dispersion, which approximates an even distribution on the network. In a first step, shortest paths were calculated from each $v' \in \mathrm{V}'$. In a second step, weights were assigned to each point of origin based on the normalized population numbers of their respective municipal area, since we assume the chance of starting a trip as well as the number of potential visitors to be higher in more densely populated areas. The process of calculating route frequency, a step which is described in Sect. 3.2, incorporates the resulting weight coefficients.

3.1.3 Points of Origin Derived from Street Network Density

A further approach involves deriving the points of origin based on the density of the street network. This approach has the advantage that no other input data are necessary while, at least to some degree, an approximation to population density distribution can be achieved. First, the polylines from the street network are used to calculate line density values and then normalized to a range from 0 to 1 on a cell-by-cell basis. The following formula is used for the normalization:

$$D_{norm} = \frac{D_i - D_{\min}}{D_{\max} - D_{\min}}.$$

Subsequently, all cells with $D_{norm} > 0$ are converted to points and projected onto the closest edge $e \in \mathrm{P}$. Thus, in this specific approach, urban agglomerations are not only favored by higher weight factors D_{norm}, but also by the fact that within these areas, a comparatively higher number of points are created (see Fig. 2e).

3.2 Calculating Route Frequency

Based on the created set of points of origins V', shortest paths are calculated for each origin-destination pair v'–v_d. Each segment on the path is assigned an individual weight coefficient, which corresponds to 1 for unweighted origins and to the weight of the origin in the weighted calculations. Route frequency $f(e_{1...n})$ is measured by calculating the number of overlapping routes per network segment, counting each route either once (following unweighted approaches) or according to its individual weight coefficient.

As has been described previously, in the case that several v' were located on one edge e, in order to minimize computational effort, the vertices are combined into one representative point. In this case, the accumulated mean weight value will determine how one route is counted when calculating the frequency.

According to the frequency interpretation of probability, one can infer the probability $p(e_{1...n})$ from the frequency $f(e_{1...n})$. Finally, both a prediction of vehicle distribution in the network and the most probable route can be deduced from the resulting network using probabilities as weights.

4 Results and Discussion

As expected, the results vary depending on the approach used. The main differences occur between the weighted and unweighted methods, whereas the contrast between the various weighted methods is comparatively small (see Fig. 3). When using the unweighted regularly-dispersed-method the calculated routes are more dispersed on the network. In comparison, the calculated routes for the weighted regularly-dispersed-method lead through the agglomeration areas (in this case the two bigger cities Aichach and Friedberg). The routes from unweighted methods also pass more often through the northern network, whereas the routes resulting from the weighted methods make more use of the southern network. However, the southern network is less frequented with the weighted 1,000 m method than with the weighted 500 m method. This distinction might make a difference in planning and managing traffic for a huge event at the stadium.

If the number of random points is too small (i.e., 50), all routes that are calculated are represented as higher frequency (compared to no frequency), whereas in the case of 1,000 random points enough routes are calculated across a segment that a meaningful frequency pattern can emerge and outliers are detected as such.

The method using 1,000 random points and the method using line density produce almost identical frequencies (see Fig. 4). This is remarkable since the line density uses about 7,000 points in contrast to the 1,000 points of the random method. An explanation might be that the randomly distributed points on the network implicitly factor in the density of the network, because they are projected onto the network after creation.

The southern network is most frequented in the 50 random, 1,000 random and line density approaches (the last southern segment is dark blue in contrast to light blue or even green in the point dispersed approaches).

Factoring in the aim to use as few additional sources as possible and to reduce computing time without losing information an approach based on deriving the potential points of origin from the street network seems most reliable. With respect to the original question of using geometric versus geographically weighted information, we have to conclude that (at least in this case study) the geometric information is sufficient to produce a satisfying route probability.

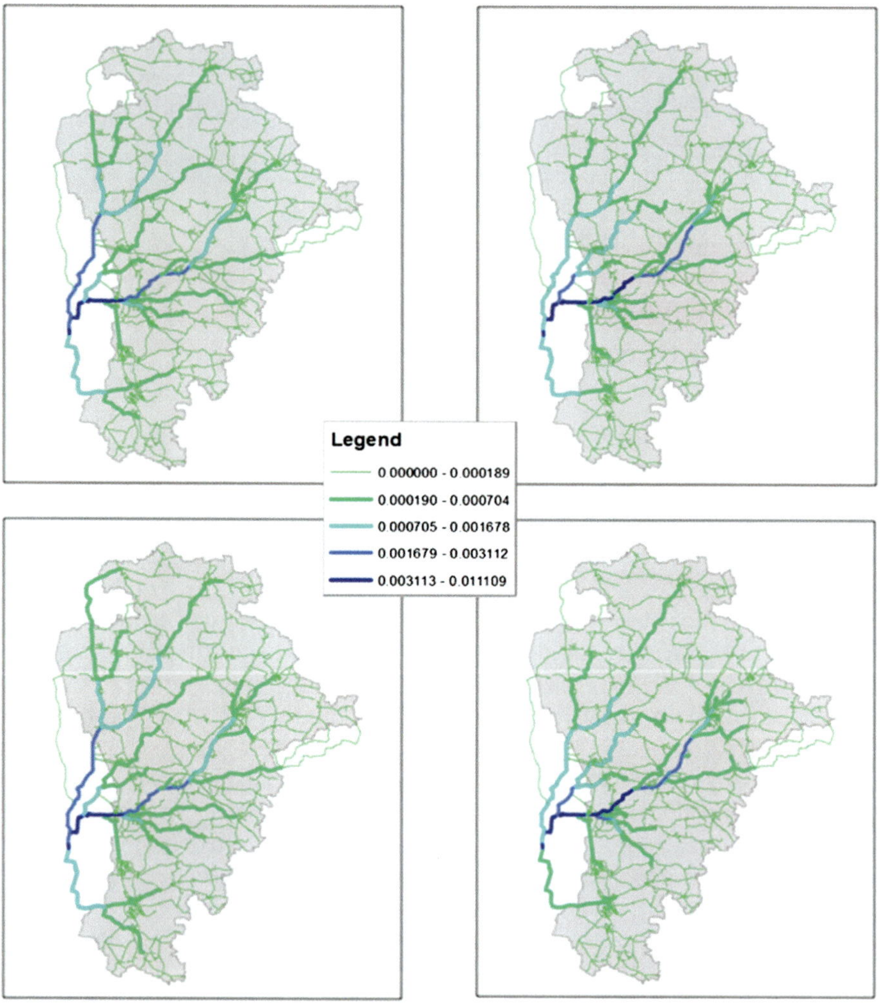

Fig. 3 Frequency of routes in the network—calculations for regularly dispersed origins. *Top row* 500 m distance, *bottom row* 1,000 m distance, *Left column* unweighted, *right column* weighted

4.1 Comparison of Frequencies

The frequencies are calculated for each segment, i.e., edge, within the network. For a comparison of frequencies per segment, a normalization procedure needs to be carried out. We have used two methods for normalization. First, we normalize across the distribution of frequency values of all segments within the network, i.e., computing a relative frequency: within calculation method 1 (e.g., 1,000 randomly distributed points) look for the maximum value $frequ_{max,method1}$ and divide the

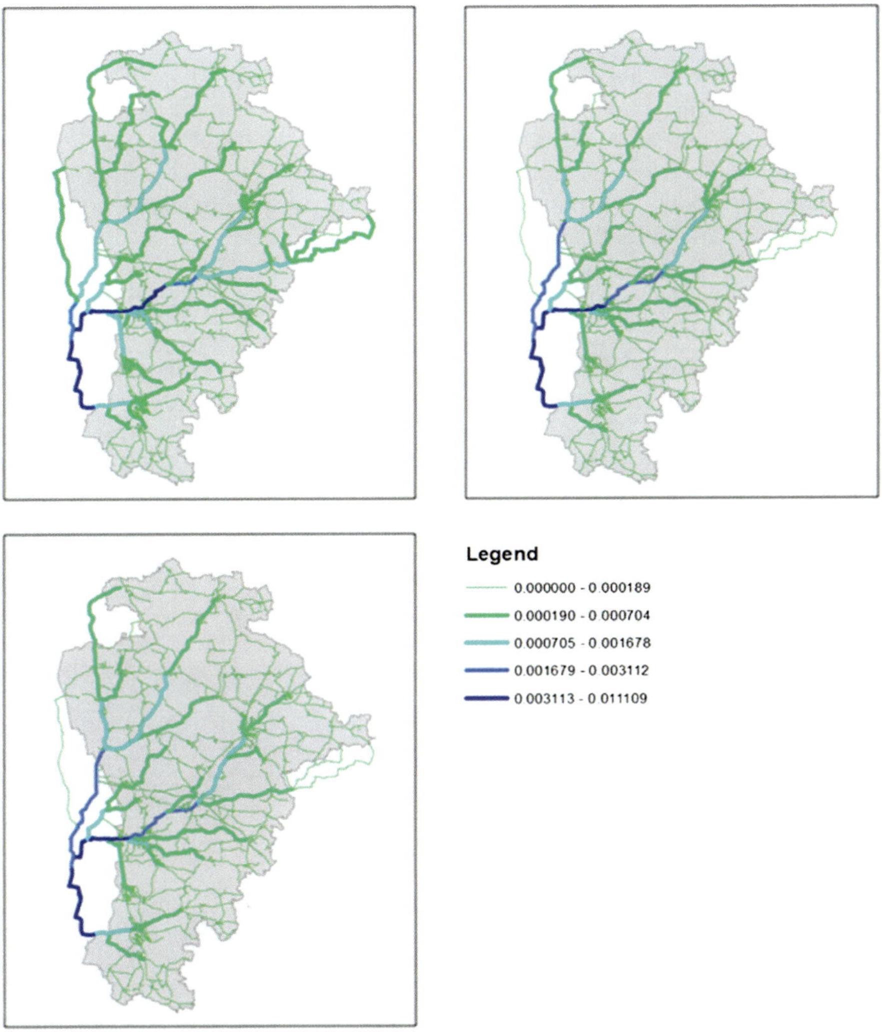

Fig. 4 Frequency of routes in the network—calculations for randomly dispersed origins (*top row left* 50, *right* 1000 points) and line density

calculated frequency $frequ_i$ of each segment i by the maximum value, thus yielding a relative frequency for each segment $relfrequ_i$ within the range 0–1. We call this relative frequency the R-value of a specific segment within a specific method.

Second, we normalize across all existing relative frequency values on one single segment for each of the seven methods used to determine the distribution of the points of origin. Thus, we divide the $relfrequ_i$ of a segment i by the maximum relative frequency across all seven methods for a specific segment i, resulting in a comparison of the relative frequency across all methods. This is only possible because the relative frequencies are normalized and thus within the range of 0–1. We call this the O-value.

In addition to the relative frequency per method (R-value) and the relative frequency per segment (O-value), a mean value of all relative frequencies as well as a standard deviation of the relative frequency can be calculated for each segment, again within the network (R-values) and across all methods (O-values). In Fig. 5 it is immediately visible that the mean relative frequency is particularly high in the last segment from the south although this segment only caters to a small portion of the total area of potential origins. In the 3D figure the last segment from the south shows a high standard deviation, which means that it is the least stable segment within the network. Since this segment carries a relatively high load, any measures of event managers or traffic planners should consider this segment very carefully.

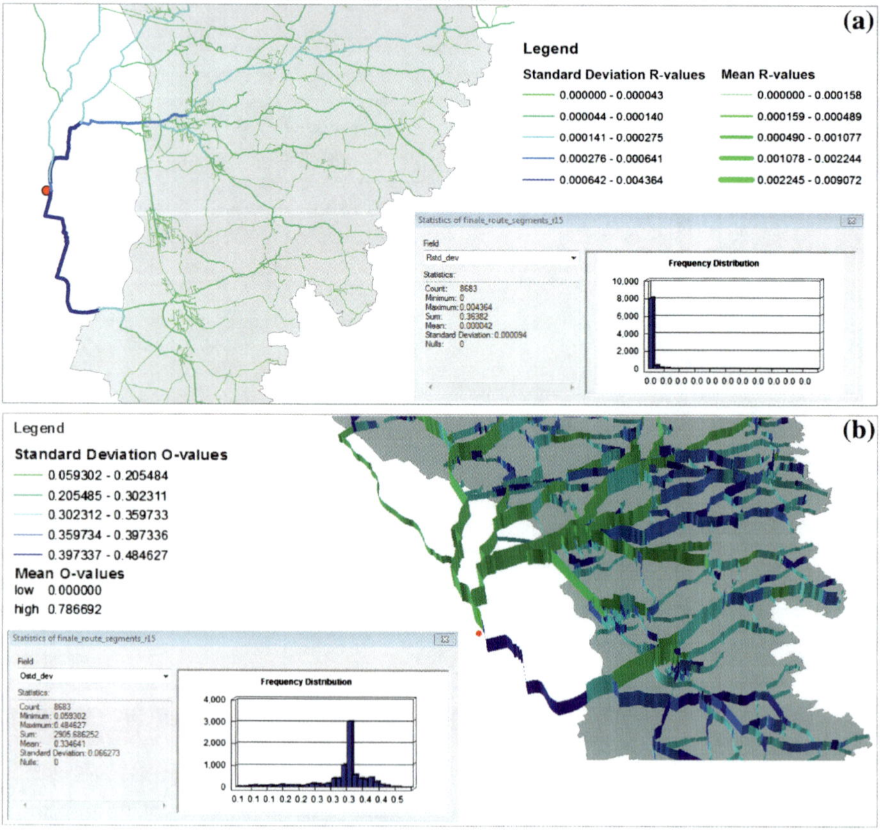

Fig. 5 Mean and standard deviation of route frequency: R-values and O-values. **a** Width of line equals mean of R-value per segment and **b** height of line equals mean of O-value per segment. *Colour* represents standard deviation

5 Conclusions and Future Work

The aim of this research was to derive the variability in the most probable routes from a specified area (represented as polygon) to a specific destination without knowing the exact points of origin. This kind of question occurs for example during mass events, when many people from different geographic areas arrive for the event. Alternatively, there could be an inability to derive the current location, e.g., due to continued missing GPS signals, while still needing to provide a route to a specified destination. Within the context of event management, event planners want to be able to provide participants with up-to-date information along the routes they most probably will take. Traffic managers would use the information on the routes most probably taken to manage traffic in order to avoid congestion or at least manage the flow of vehicles.

The approach taken here differs from traditional route assignment since it derives a set of potential points of origin through several methods and calculates a set of shortest routes starting from these points of origin. Each of the segments within the road network can then be attributed with the number of routes passing through the segment. This number is then used to calculate a probability of the segment being part of a route to the destination starting from within the specified area of origin.

Our results show that the differences between the methods in determining the points of origin produce only slightly different probabilities. Considering that the calculation of a potentially high number of shortest paths takes up resources, we chose as the best the method with the least amount of necessary calculations. For this case study at least it turns out, that the geometric information, i.e., the road network itself, is sufficient to generate the probability values for route segments.

Taking the results a step further, we could determine the most probable route through the whole road network, starting from the destination v_d and using the calculated route segment probabilities. In this case a breadth first search needs to be conducted which allocates probability values to each edge. The result shows the most probable route from an area to the destination.

What should be done in the near future is the comparison of the probabilities of the route segments with centrality network measures as well as results of TAZ centroid-based routing processes at a number of structurally different networks. This would correlate the geometric properties of the road network structure with the usage properties as well as test the comparability of methods. The results could be validated for example by deriving the observed frequency of use of a specific road segment from floating car data or traffic counts obtained from road-side traffic counters.

Additional variations in the calculations performed in this study may be enlightening: a distinction between road types in the shortest path (i.e., hierarchical shortest path) or using different path optimizations (fastest, most beautiful, least complex…) could produce a different result in the final probabilities. Another distinct approach could be the use of a probability surface instead of a distinct polygon in the delineation of the origin area. Future work could also encompass the derivation of the most central route through the network, resulting in a centrality measure for routes within the network. Another variation of the computation could incorporate temporal measures instead of distance measures for the calculation of the relative frequencies.

Acknowledgements We gratefully acknowledge support of Carolin von Groote-Bidlingmaier through the program "Chancengleichheit von Frauen in Forschung und Lehre" of the University of Augsburg.

References

Basiri A, Winstanley A, Sester M, Amirian P, Kuntzsch C (2012) Uncertainty handling in navigation services using rough and fuzzy set theory. In: Kroeger P, Renz M (eds) QUeST '12 Proceedings of the Third ACM SIGSPATIAL international workshop on querying and mining uncertain spatio-temporal data. Redondo Beach, CA, USA, 07 Nov 2012

Chang KT, Khatib Z, Ou y (2002) Effects of zoning structure and network detail on traffic demand modelling. Environ Plan 29:37–52

Gonzales JP, Stentz A (2007) Planning with uncertainty in position using high-resolution maps. In: Proceedings IEEE international conference on robotics and automation, Rome, Italy, 2007

Goodchild MF (2009) Methods: uncertainty. In: Kitchin R, Thrift M (eds) International encyclopedia of human geography. Springer, New York

Hait A, Simeon T, Taix M (1999) Robust motion planning for rough terrain navigation. In: Proceedings. IEEE/RSJ International. Conference. Robotics and Systems, Kyongu, Korea

Hajek A (2012) Interpretations of Probability. In: Zalta EN (ed) The stanford encyclopedia of philosophy. http://plato.stanford.edu/entries/probability-interpret/#FreInt. Accessed 30 May 2014

Hazewinkel M (2001) Law of large numbers. In: Hazewinkel M (ed) Encyclopaedia of mathematics. Springer, Berlin

Jaynes ET (2003) Probability theory—the logic of science. Cambridge University Press, Cambridge

Liao F, Rasouli S, Timmermans H (2014) Incorporating activity-travel time uncertainty and stochastic space-time prisms in multistate supernetworks for activity-travel scheduling. Int J Geogr Inf Sci 28(5):928–945

MacNally MG (2007) The four step model. In: Hensher DA, Button KJ (eds) Handbook of transport modeling. Elsevier, Oxford

Miller HJ, Shaw S-L (2001) Geographic information systems for transportation: principles and applications. Oxford University Press, Oxford

Prato CG (2009) Route choice modelling: past, present and future research directions. J Choice Modeling 2(1):65–100

Qian ZS, Zhang HM (2012) On centroid connectors in static traffic assignment: their effects on flow patterns and how to optimize their selections. Transp Res Part B 46:1489–1503

Qiu D, Papotti P, Blanco L (2013) Future locations prediction with uncertain data. In: Blockeel H, Kersting K, Nijssen S, Zelezny F (eds) Machine learning and knowledge discovery in databases, LNCS 8188. Springer, Berlin, pp 417–432

Wang S, Wenzhong S, Yuan H, Chen G (2005) Attribute uncertainty in GIS data. Fuzzy Syst Knowl Discov, LNCS 3614:614–623

Worboys M (1998) Imprecision in finite resolution spatial data. Geoinformatica 2(3):257–279

Zhang J, Goodchild MF (2002) Uncertainty in geographical information. Taylor and Francis, New York

Visualization and Communication of Indoor Routing Information

Jukka M. Krisp, Mathias Jahnke, Hao Lyu and Florian Fackler

Abstract In this paper we investigate the display and communication of indoor routing instructions via small maps, map-like graphics and non-photorealistic presentations of interior spaces. Our goal is to elaborate on questions like, "how can we provide an optimal depiction of the navigation information so that the user is able to find the destination within an indoor environment? The destination could be for example a classroom or a particular office in a university environment. We provide a case study for indoor routing maps within the Technical University Munich's main building. Currently available data on floor footprints, points of interests (POIs), indoor-landmarks and the routing graph for the building are used to implement a routing service. The routing information is displayed on a smart phone implemented in the Windows 7.8 operating system. Designs of different ways to display routing instructions on smart phone displays are investigated. Experiences for building up the map design for this particular indoor routing system can be transferred to other buildings and provide a basis for the design ideas of indoor routing maps.

Keywords Routing · Indoor navigation · Way-finding · Location-based services (LBS) · Mobile maps

J.M. Krisp (✉)
Department of Geography, Universität Augsburg, Alter Postweg 118,
86159 Augsburg, Germany
e-mail: jukka.krisp@geo.uni-augsburg.de

M. Jahnke · H. Lyu · F. Fackler
Department of Cartography, Technische Universität München, Arcisstraße 21,
80333 Munich, Germany
e-mail: mathias.jahnke@tum.de

H. Lyu
e-mail: Hao.Lyu@lrz.tum.de

F. Fackler
e-mail: flofackler@msn.com

G. Gartner and H. Huang (eds.), *Progress in Location-Based Services 2014*,
Lecture Notes in Geoinformation and Cartography,
DOI 10.1007/978-3-319-11879-6_3

1 Introduction

The growing amount of smart phones and tablet-computers the use of indoor navigation functionalities on these devices is becoming increasingly popular. This results in a growing application and research field of Location-Based Services (LBS). LBS are investigated from different perspectives including mobile positioning and tracking technologies, data capturing and computing devices, integrated software engineering, and user studies for various applications (Krisp 2013). Currently many researchers and companies examine and develop applications to provide users with navigation services for indoor environments. In many cases, these services build up and use existing technological approaches from in-car or outdoor navigation systems.

Indoor navigation is one of the prominent research areas related to LBS. Indoor navigation can be separated into a number of sub research areas. These cover indoor positioning, which includes research on positioning via WLAN, Bluetooth (Nguyen and Luo 2013; Naya et al. 2005), earth magnetic field (Bilke and Sieck 2013) and other sensor related positioning technology, like RFIDs (Romanovas et al. 2013). Indoor navigation data and data formats need standardization which are currently under development (Donaubauer et al. 2013; Li et al. 2013). In addition a wide field of research includes algorithms and computation, including computational performance, and for example travel mode classifications (Zhang et al. 2013). One of the key researches is visualization and communication of routing information.

In this paper we focus on the visualization and communication of indoor navigation information. There are currently no standards or best practices of how to display indoor navigation information. Several attempts use two-dimensional floor plans or three-dimensional building models and plot the current position, plus the routing information. There seems to be no formal investigation, if these representations are understood by the users and help the users in their navigational tasks.

New technical devices like smart phones with large touch-screens, head-mounted devices (like Google Glass) or "smart watches" offer new ways to communicate indoor navigation information. Also more traditional devices like paper maps of floor plans may have advantages as Lorenz et al. (2013) state, "both, paper and mobile devices have their advantages and disadvantages regarding their suitability as presentation medium for personal indoor navigation maps [...]. Provided that map design follows cartographic principles, they are most suitable for guiding a user to a desired destination. [...] one of the biggest challenges will be to find out which indoor configurations influence user needs regarding map design". (Lorenz et al. 2013).

2 The Indoor Data Challenge

Indoor data modeling is an essential part of providing indoor navigation service. The nested and enclosed structure of a building that is different from typical outdoor environment makes it infeasible to directly use outdoor data modeling methods. We identified a number challenges concerning data, to provide "good" indoor navigation services. These include the organization of data and how can we acquire indoor data. Additionally a lack of a formalized indoor data model also increases the difficulty to evaluate the usability of a certain visualization method under certain contexts. Early approaches of indoor navigation have already tried various visualization methods like traditional 2D maps, 3D visualization and augmented reality (Abowd 1997; Höllerer 1999) while how to modeling these data is not fully explored.

Computer-Aided Design (CAD) files are widely used in architecture and able to provide geometric representation with concept of building elements, thus become important data source for indoor navigation systems. However, based on a CAD file format, an indoor navigation system's function is very limited due to the lack of deeper indoor knowledge of such data format. The knowledge includes topological relationships between indoor objects. Convert CAD files into Geographic Information System (GIS) file format seems to be able to overcome this shortage. Examples are Orientation Tool of École Polytechnique Fédérale de Lausanne (EPFL) (Gilliéron and Merminod 2003), BMW Personal Navigator (BPN) (Krüger et al. 2004) and Indoor/Outdoor Mobile Navigation System (Nikander et al. 2013). Since Volunteered Geographic Information (VGI) and crowdsourced geodata acquisition have been utilized successfully in outdoor applications, their possibilities of providing indoor navigation are also studied (Goetz and Zipf 2013).

Although 2D planar data is widely used for visualization and communication (Meijers et al. 2005), the need of 3D information is increasing. Researches have been done to explore the suitability of existing data formats for indoor navigation. As investigated by a number of researchers, Hijazi et al. (2011) and Hijazi and Ehlers (2009) indoor data can be modeled stored in Keyhole Markup Language (KML) format which makes it easy to be visualized in Google Earth. Existing specialized software like 3D Max and Sketchup are capable to depict complex geometry of indoor objects, while the data formats are insufficient to preserve semantic information. Building Information Model (BIM) can provide sufficient 3D information about a building itself and Industry Foundation Classes (IFC) provides a good reference for BIM. BIM usually contains detailed geometric and semantic information though out a facility's whole life cycle. However, due to its complexity and lack of topological relationships, BIM itself is not feasible to be indoor navigation data model. A middle model or a BIM based model may be more suitable for indoor navigation task. Isikdag et al. (2013) proposed a BIM Oriented Modeling methodology resulting in a new BIM based model (BO-IDM) for indoor navigation. City Geography Markup Language (CityGML) which is an Open Geospatial Consortium (OGC) standard about representation, storage, and exchange of virtual 3D city and landscape models also includes a description of indoor environment. CityGML provides geometric, topologic and semantic information for city object modeling. In Level of Detail 4 (LoD4) the interior of a building is represented in the building model. Several mechanisms such as Application Domain Extension (ADE) ensure CityGML to be extended to satisfy special applications. In Becker et al. (2009a, b) a Multilayered Space-Event Model is proposed for indoor navigation.

The model introduces a multi-layer structure to integrate different conceptual division according to different subtasks in indoor navigation as positioning, route planning and route communication on both (Euclidean) geometric and topologic space. This model can be implemented by extending CityGML. Additionally the OGC is working on an IndoorGML standard (Li et al. 2013) within a working group. The extended CityGML and IndoorGML seems to be a promising solution that can partially solve the indoor data problems, but we hope to see more practical indoor navigation applications to illustrate their usability.

3 Challenges of Visualizing Spatial Indoor Routing Data in a 3D Environment

The representation of 3D information is an evolving area and a challenging task in cartography. 3D information in particular in the field of city models ranges from block models to very detailed architectural models. The indoor information can be integrated using CityGML (LOD4) as mentioned before. After integrating this extra information into a city model, the next step is to visualize the interior information and communicate them to a distinct user in a feasible and usable way. Beside other promising approaches of visualizing 3D data, the non-photorealistic approach (Strothotte and Schlechtweg 2002) seems to be an outstanding one which stands for an aesthetic appeal (Plesa and Cartwright 2008) and gives more degrees of freedom to include semantic information into the visualization. The non-photorealistic visualization originates from computer graphics and was first used for technical illustrations (Gooch et al. 1998).

Visualizing 3D spatial information is not only a rendering task of complex geometries. For a user centered visualization cognitive (Swienty et al. 2008) and usability (Jahnke 2013) aspects have to be taken into account. Gestalt laws and the knowledge about spatial perception (Gerrig and Zimbardo 2008) are the basis for communicating 3D spatial information to the user.

A popular approach of integrating indoor building plans into maps can be found at Google Maps™ which gives a first impression of bridging the gap between outdoor and indoor information. Nevertheless research has to be performed on the question of integrating and communicating indoor routing information in a 3D environment. In particular in the 3D environment the joint visualization of the routing information and the building geometry is a challenging task and has to be adjusted to each other when represented together.

Challenges which occur in particular when visualizing 3D indoor models are that only the inner space of the building can be used for navigating and in most cases the view along the aisle of a floor gives the user the information for orienting themselves. Therefore, it is a demanding visualization task to give the user an overview of the entire route in particular if the route is crossing different floor levels. Here the visualization can become slightly confusing if not well designed. Therefore, occlusion seems to be a major problem but it can be avoided using transparency to look inside and through walls or buildings. A drawback of using transparency within a 3D visualization consists of many features that will become visible either important or unimportant. The occurring unimportant features disturb the user in getting the needed information as well as they make the whole visualization hard to understand. Therefore, we choose small maps and map-like graphics together with representations of interior space to display and communicate indoor routing instructions as a basis for this work.

4 Case Study—Indoor Navigation in the Technical University Munich

To provide a design for indoor routing maps for a case study we selected the Technical University Munich's main building. Within this environment we offer a possible solution to questions like: how can we provide an optimal depiction of the navigation information so that the user is able to find the destination within an indoor environment? Currently available data on floor footprints, points of interests (POIs), indoor-"landmarks" and the routing graph for the building are used to implement an indoor navigation service. The navigations service provides the user a route from an origin to a destination using the before mentioned small maps, map-like graphics and the interior information. The routing and building information are displayed together on a smart phone.

The available data seems to be the bottleneck in these indoor navigation services, as data is lacking a machine-readable file format containing doors, rooms, entrance areas, as well as semantic information like room numbers and level information. In particular semantic information like the room numbers attached to the 3D features are missing in general within the data we got for this case study. For this case study the indoor information was provided by the university administration with two-dimensional floor plans in the pdf file format (see Fig. 1). The provided indoor plan contains information such as pillars, walls, windows, doors, stairs and hatches.

To acquire consistent polygons for floors, rooms and doors out of the provided pdf-file the unnecessary information has to be deleted by using a vector graphics editor like Adobe Illustrator™ or Inkscape. The polygons representing floors, rooms or doors have to be modeled in different layers to better differentiate between those objects. During the modelling phase all polygons have to be properly aligned to each other for topological correctness. The drawback of the workflow we used for this case study is the manual work which has to be accomplished to get a correct aligned and machine-readable 2D floor plan for the indoor environment. At least the

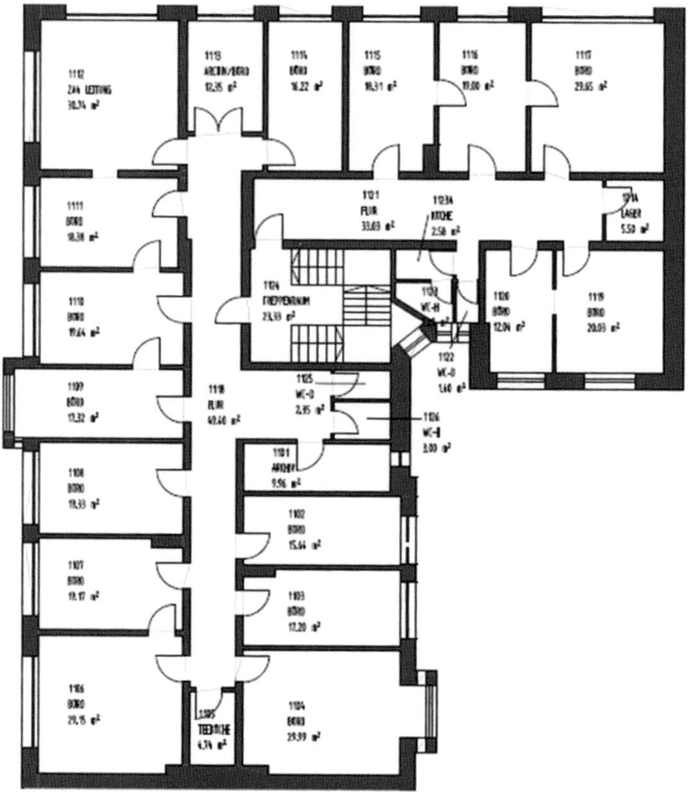

Fig. 1 Illustration of a two-dimensional map representing an indoor plan

room numbers have to be attached to the features representing the rooms. The last step in this workflow is to derive 3D features out of the 2D floor plan by extruding all features to a distinct height. As well stairs and elevators have to be modelled in 3D to connect floors.

In this particular case study we have chosen the students as our main user group. In our case the student is a slightly experienced user in terms of knowing the university environment as well as in using 3D visualization on a smartphone. Therefore the visualization should extend the users mental model of the environment and should help to reach the destination location. Beside this user group university visitors as well as emergency services are possible user groups for an indoor navigation service.

To create a 3D indoor model was the aim of our design. In addition to basic elements such as room numbers, entrance areas to the university and stairs, more supplementary information is indicated. These supplementary information include: student service center, library, cafeterias, canteen card recharge stations and restrooms. Figure 2a, b show the base model of the indoor environment.

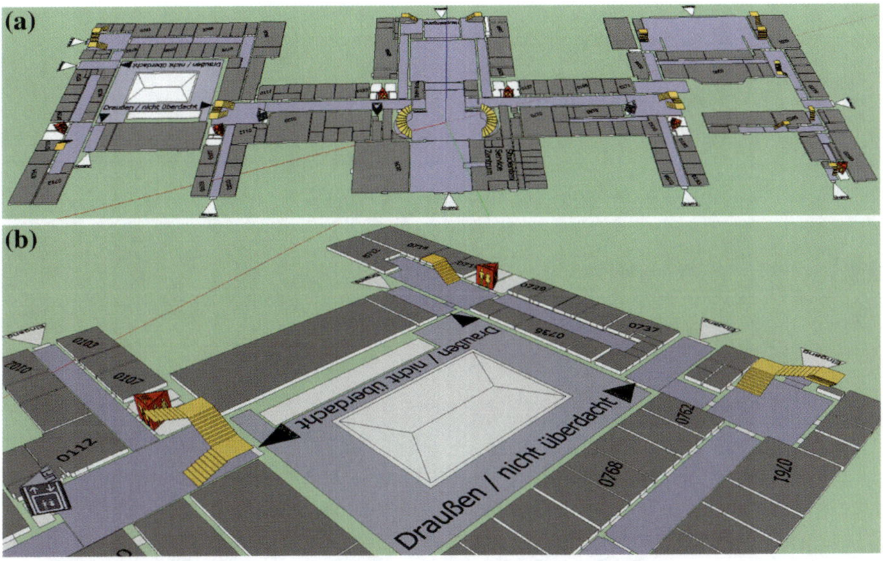

Fig. 2 **a** Overview map of the ground floor, **b** detailed map view

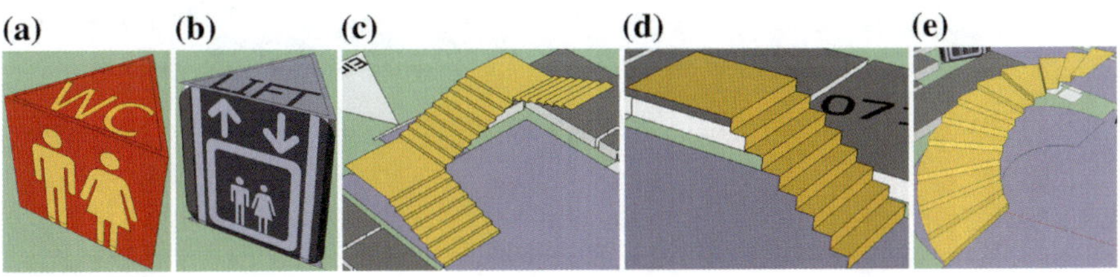

Fig. 3 Details for specific depictions of potential "indoor landmarks" within a 3D navigation system, **a** wc, **b** lift, **c–f** pre-defined shapes for different types of stairs

As shown in Fig. 2, the room numbers are shown at the beginning and the end of each corridor, to avoid an information overflow in the visualization. The room numbers are not perfectly aligned to the corridors due to better fit into the polygons representing the rooms and to better attract the users attention to recognize the room numbers within the visualization. Additionally all room numbers are displayed if the user reaches a specific zoom level.

Depicting specific indoor landmarks or POIs seems to be crucial for the user to complete the navigation task. Therefore a number of significant POIs are included. Figure 3a, b shows the created symbols for toilets and elevators. They have a triangular shape to be clearly visible from any direction. The pictograms are widely used and the meaning of them is clear. Depending on the local situation any of the variations of stairs, shown in Fig. 3c–f, are displayed.

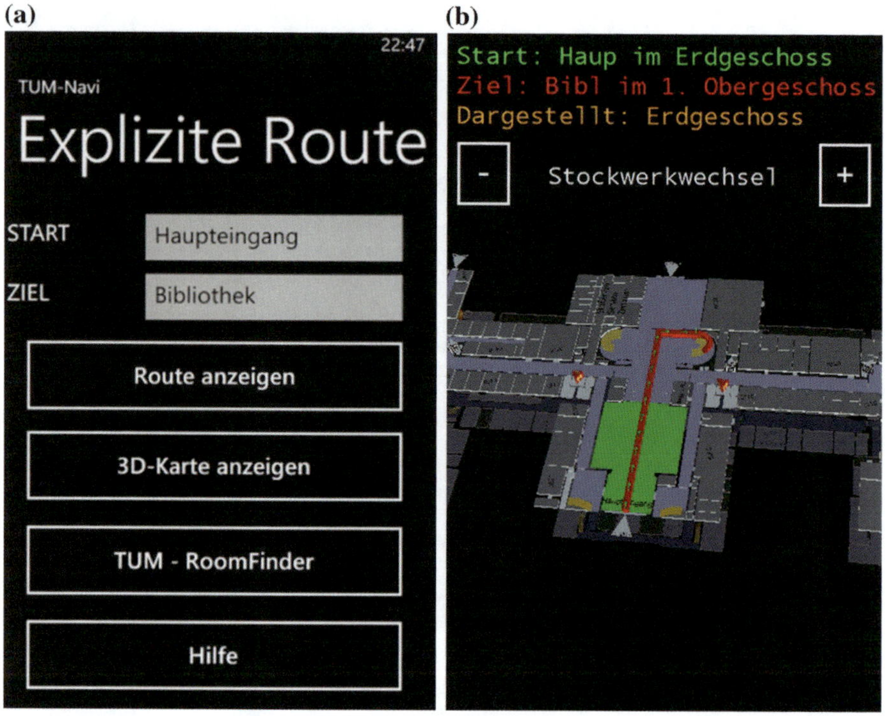

Fig. 4 **a** User interface, **b** map view of the case study application for the Technical University Munich indoor routing system

After starting the application the start screen (Fig. 4a) is shown and the user starts routing by entering the start (START) and destination (ZIEL) the particular route are displayed on an overview base map of the specific floor (Fig. 4b). Figure 4a shows a routing from the main entrance (Haupteingang) to the library (Bibliothek) while Fig. 4b shows the route (red line) from the main entrance to the stairs. In this case, the stairs connect the basement with the first floor on which the library is located.

The textual information in Fig. 4b contains the starting area colored in green and the destination area colored in red. Additionally it displays which level can be seen on the screen. With the minus and plus buttons the level can be changed from basement to the first floor. Therefore, the user gets the information from where to start, where he can find his destination and which level is shown on screen.

To start navigation the user can zoom into this area and will recognize the red route with the arrows on top, which indicate the walking direction. It is possible to show the direction, because the map is always aligned towards north. This utilizes the smartphones build-in gyroscope. After receiving the measured rotation angles for the three axes x, y and z, the rotation matrices are calculated and based on this the model is rendered again. Though the 3D-model of the university is always

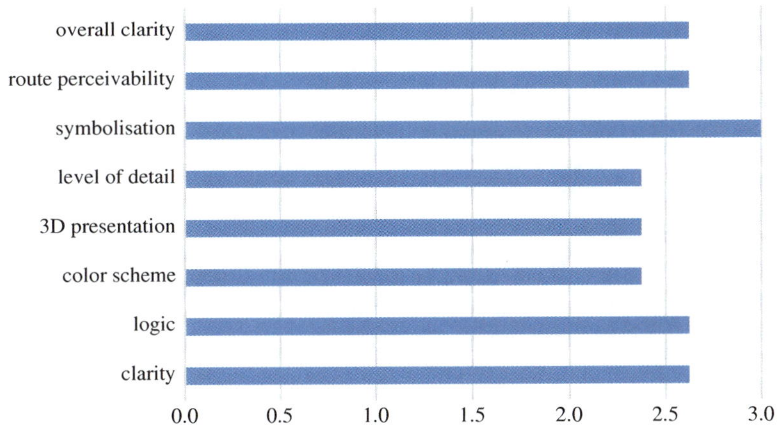

Fig. 5 Selected questions and its average results from the initial user survey

correctly oriented, thus objects in the visualization are always oriented according to how they appear in real world. This makes the mapping from the visualization to the real world easier and less cognitive demanding for the user.

The sample application is implemented on a Windows Phone running the Windows Phone 7.8 operating system. To examine the usability of our approach we conducted a first survey based on a user feedback form with students from Technical University Munich. The students were equipped with a mobile phone running the navigation application and they have to fulfill different routing tasks. Afterwards they filled in the provided feedback form. Of major interest was the first impression of the potential users and if they were able to understand the visualization and to follow the displayed routing information within the university environment.

The survey is based on a small group of eight students to obtain first insights if and how the visualization approach works. The survey shows that the users were quite familiar with the university environment as well as with the handling of the 3D model on a mobile phone. The users had to answer questions like how they estimate route perceivability as well as how useful the 3D representation or the overall clarity is. The answers to these questions have to be ranked on a scale ranging from −3 (very bad) to +3 (very good).

Figure 5 shows selected questions and their ranking. It shows that that the visualization approach seems to attract the user. In particular, the overall clarity of the application, the route perceivability, the logic and clarity as well as the symbolization are judged as very good. Therefore, our developed visualization approach seems to lead the users reliably to their destination. Nevertheless, the approach needs more investigations using a more detailed survey based on a broader user basis. Additionally the design has to meet cartographic principles and theories.

5 Conclusion and Discussion—Creating Indoor Navigation Systems

A number of questions arise based on the case study in this paper. These can be organized into three major challenges that need to be considered when building an indoor navigation system:

- *The data challenge*
  Where and to what LOD can data be acquired? Additional research may reveal if we can utilize VGI based data to build up a database that is suitable for indoor navigation services? How can we consider real-time data streams (for example positions of other persons), in addition to static data (like indoor floor plans)?
- *The route computation challenge*
  How can computational methods be applied to receive the desired outcome? How can we provide the user with a "natural route" (Krisp and Liu 2009) of walking indoors? Preferably that would be a similar route that the user would take, if he/she would know the building very well?
- *The visualization and communication challenge*
  How can the indoor navigation system display the route that the user can find the way? Do we need device specific designs? For example different communication of the information on smart-phones, Google glass, smart watches (for example the Galaxy Gear)?

The case study reveals that the 3D visualizations created meet the expectations of users. The users pointed out that by using a system, which is automatically aligned, it can be avoided that users are walking into a wrong direction. The representation of a route is clear and can easily be understood. Users prefer a clear overview of the route and are interested in seeing their actual position.

In the case study implementation of the indoor navigation system for the Technical University Munich, the user has to push a "−" and "+" button when changing the floor. This solution seems to be insufficient as many users did not understand the use of these buttons. When using data models in indoor navigation systems, we should consider cartographic methodologies to produce a better visualization result for users. This is a demanding task by means of cartographic principles.

A challenge and something to avoid in the data collection is manual data acquisition. Data collection and processing should be automated. The automation of deriving 3D indoor building models is a big issue in further research and has to be investigated in the future. Additionally the map design for particular indoor routing systems can be transferred to other systems and provide a basis for the design parameters of indoor routing maps.

Acknowledgments We thank the university administration of Technical University Munich (TUM) to provide us the pdf floor plans and we are grateful of the work that Mr. Bogdan Cirlugea did in means of creating a sufficient number of floor plans for testing purposes.

References

Abowd GD (1997) Cyberguide: a mobile context-aware tour guide. Wireless Netw 3:421–433

Becker T, Nagel C, Kolbe T (2009a) A multilayered space-event model for navigation in indoor spaces. 3D geo-information sciences. Springer, Berlin

Becker T, Nagel C, Kolbe T (2009b) Supporting contexts for indoor navigation using a multilayered space model. In: 10th international conference on mobile data management systems, services and middleware. IEEE, pp 680–685

Bilke A, Sieck J (2013) Using the magnetic field for indoor localisation on a mobile phone. In: Krisp JM (ed) Progress in location-based services. Springer, Heidelberg

Donaubauer A, Straub F, Panchaud N, Vessaz C (2013) A 3D indoor routing service with 2d visualization based on the multi-layered space-event model. In: Krisp JM (ed) Progress in location-based services. Springer, Berlin

Gerrig RJ, Zimbardo PG (2008) Psychologie. Pearson, München

Gilliéron P-Y, Merminod B (2003) Personal navigation system for indoor applications. In: 11th IAIN world congress, Berlin, Oct 21–24

Goetz M, Zipf A (2013) Indoor route planning with volunteered geographic information on a (mobile) web-based platform. In: Krisp JM (ed) Progress in location-based services. Springer, Berlin

Gooch A, Gooch B, Shirley P, Cohen E (1998) Non-photorealistic lighting model for automatic technical illustration. In: Proceedings of the 25th annual conference on computer graphics and interactive techniques. ACM, New York, pp 447–452

Hijazi I, Ehlers M (2009) Web 3D routing in and between buildings. Fourth national GIS symposium, Al-Khobar

Hijazi I, Zlatanova S, Ehlers M (2011) NIBU: an integrated framework for representing the relation among building structure and interior utilities in micro-scale environment. Geo spat inf Sci 14:98–108

Höllerer T (1999) Exploring MARS: developing indoor and outdoor user interfaces to a mobile augmented reality system. Comput Graph 23:779–785

Isikdag U, Zlatanova S, Underwood J (2013) A BIM-oriented model for supporting indoor navigation requirements. Comput Environ Urban Syst 41:112–123

Jahnke M (2013) Nicht-photorealismus in der stadtmodellvisualisierung für mobile nutzungskontexte. Technische Universität München

Krisp JM (2013) Progress in location based services (LBS). Springer, Berlin

Krisp JM, LIU L (2009) Pedestrian off-road navigation for multi modal routing. In: 6th international symposium on lbs and telecartography. Nottingham, UK

Krüger A, Butz A, Müller C, Stahl C, Wasinger R, Steinberg KE, Dirschl A (2004) The connected user interface: realizing a personal situated navigation service. In: Proceedings of the 9th international conference on intelligent user interfaces, ACM, New York, pp 161–168

Li M, Goetz M, Fan H, Zipf A (2013) Adapting OSM-3D to the mobile world: challenges and potentials. In: Krisp JM (ed) Progress in location-based services. Springer, Berlin

Lorenz A, Thierbach C, Baur N, Kolbe TH (2013) App-free zone: paper maps as alternative to electronic indoor navigation aids and their empirical evaluation with large user bases. In: Krisp JM (ed) Progress in location-based services. Springer, Heidelberg

Meijers M, Zlatanova S, Pfeifer N (2005) 3D geoinformation indoors: structuring for evacuation. In: Proceedings of next generation 3D city models, pp 21–22

Naya F, Noma H, Ohmura R, Kogure K (2005) Bluetooth-based indoor proximity sensing for nursing context awareness. In: Proceedings of the 9th IEEE international symposium on wearable computers, pp 212–213

Nguyen K, Luo Z (2013) Evaluation of bluetooth properties for indoor localisation. In: Krisp JM (ed) Progress in location-based services. Springer, Heidelberg

Nikander J, Järvi J, Usman M, Virrantaus K (2013) Indoor and outdoor mobile navigation by using a combination of floor plans and street maps. In: Krisp JM (ed) Progress in location-based services. Springer, Berlin

Plesa MA, Cartwright W (2008) Evaluating the effectiveness of non-realistic 3D maps for navigation with mobile devices. In: Meng L, Zipf A, Winter S (eds) Map-based mobile services: design, interaction and usability

Romanovas M, Goridko V, Klingbeil L, Bourouah M, Al-Jawad A (2013) Pedestrian indoor localization using foot mounted inertial sensors in combination with a magnetometer, a barometer and RFID. In: Krisp JM (ed) Progress in location-based services. Springer, Heidelberg

Strothotte T, Schlechtweg S (2002) Non-photorealistic computer graphics: modeling, rendering and animation. Morgan Kaufmann, USA

Swienty O, Reichenbacher T, Reppermund S, Zihl J (2008) The role of relevance and cognition in attention-guiding geovisualisation. Cartographic J 45:227–238

Zhang L, Dalyot S, Sester M (2013) Travel-mode classification for optimizing vehicular travel route planning. In: Krisp JM (ed) Progress in location-based services. Springer, Berlin

A Computational Method for Indoor Landmark Extraction

Hao Lyu, Zhonghai Yu and Liqiu Meng

Abstract Researches on spatial cognition have shown human often need landmarks for an easy and successful wayfinding. For indoor navigation systems, landmark information is also important to improve their usability. However relatively little effort has been devoted to indoor landmarks from either theoretical side or practical side. In this paper we introduced several relevant theories, i.e. spatial cognition, affordance theory and space syntax, as basis to elaborate visual, semantic and structural salience indicators. The proposed indicators are used to form a computational indoor landmark extraction method. The feasibility of this method is proved by a case study.

Keywords Affordance · Space syntax · Indoor landmark extraction

1 Introduction

With the continuous expansion of urban area, more and more large and complex buildings emerge. Some of them look magnificent and embody the glorious success in civil engineering and architectural art, but may be a nightmare for people who get lost in them, particularly in case of emergency.

In fact, wayfinding in unfamiliar buildings is never an easy task. During wayfinding, people need to build up mental maps, reason a route plan, re-orientate and keep the right direction. They may get lost, whenever a mistake is made. In indoor

H. Lyu (✉) · L. Meng
Department of Cartography, Technische Universität München, Arcisstraße 21, 80333 Munich, Germany
e-mail: Hao.Lyu@lrz.tum.de

L. Meng
e-mail: liqiu.meng@bv.tum.de

Z. Yu
School of Resources and Environmental Science, Wuhan University, Luoyu Road 129, 430079 Wuhan, China

G. Gartner and H. Huang (eds.), *Progress in Location-Based Services 2014*, Lecture Notes in Geoinformation and Cartography,
DOI 10.1007/978-3-319-11879-6_4

environments, few traditional landmarks (as in outdoor environment) can be perceived; people in buildings can hardly maintain a sense of global orientation; they also tend to assume the same configuration on different floors, and get confused when floors change (Wang and Brockmole 2003; Hölscher et al. 2006). Moreover, unconventional architectural design, poorly designed sign system and numbering system also increase the chance of making mistakes.

Navigation systems with automatic routing function seems to be helpful assistants for wayfinders. However providing merely distance description without enriched information is not sufficient and sometimes even negatively influences users. Landmarks are needed to increase the usability of such navigation systems (Burnett 1998). Hölscher et al. (2006) found that checking landmarks and getting guidance from external tools can be also an important assistance in indoor wayfinding.

Many landmark based indoor navigation applications have been reported, but most of them use pre-defined landmarks or hard coded landmarks, only a few of them considered landmark selection process. In this paper we attempt to enrich the landmark concept with theories on cognition, affordance and space syntax by proposing several salience indicators and a computational method for the extraction of indoor landmarks from a geo-database.

The remaining of the paper is organized as follows: Sect. 2 gives a brief introduction of human wayfinding, navigation and landmark with a conclusion that indoor landmark extraction is still on its early stage. Section 3 dedicated to three relevant theories as basis for indoor landmark extraction. Section 4 presents measurements for individual salience indicators and a combination of these indicators for an outlier detection method to find most possible landmarks. At the end of this section, a case study demonstrates the capability of this method. The last section gives out conclusions and directions for future work.

2 Wayfinding, Navigation and Landmark

2.1 The Basic Theory of Navigation

Navigation consists of two fundamental tasks: wayfinding and motion. These two tasks are performed iteratively until people reach their destinations. The term 'wayfinding' was originally introduced by Kevin Lynch in 1960s. Golledge (1999) described it as a purposive, directed and motivated activity of determining and following a path or route between an origin and a destination. Positioning happens when people need to check their locations. Mental maps or external tools as GPS receiver or maps are often involved in positioning. (Re-)orientation happens when people face choices between different directions (or routes). Landmarks are important in both tasks and will be discussed in the following section.

Spatial knowledge for wayfinding can be categorized into three levels, i.e. landmark knowledge, route knowledge and survey knowledge. Golledge (1999) proposed that the basic geometry of space and its cognitive maps can be summarized in terms of points, lines, areas, and surfaces. Thus wayfinding environment is usually represented as route network and referenced with landmarks.

2.2 Landmark and Navigation

Lynch (1960) defined landmarks as external points of reference that are not part of a route. Presson and Montello (1988) proposed two kinds of landmarks based on wayfinding activity, i.e. route decision landmarks, which are cues for turnings at decision points; route maintenance points, which help to keep human on the routes. Lovelace et al. (1999) distinguished between landmarks at decision points (where are-orientation is needed) and at potential decision points (where a re-orientation would be possible, but should not be done to follow the current route), route marks (confirming to be on the right way), and distant landmarks. As Sorrows and Hirtle (1999) summarized, landmarks serve multiple purposes in wayfinding as organizing concept which help people build up structured spatial knowledge representation and navigational tools.

2.3 Landmark Extraction

Many researches have been reported on outdoor landmark extraction methods in recent years (Raubal and Winter 2002; Elias 2003; Nothegger et al. 2004a, b; Klippel and Winter 2005). By contrast, indoor landmark extraction is still on its early age. Millonig and Schechtner (2007) emphasized the importance of visibility for identifying indoor landmarks, and combined individual moving tracks and questionnaires to select indoor landmarks. This paper has shown that automatic selection of indoor landmark is still on its early stage and a quality-measurement system is missing.

3 Towards Indoor Landmark Theory

In this section we first describe a two-tiered cognitive model for indoor wayfinding. Then a representation of indoor places is proposed. At last space syntax theory is introduced to investigate visual salience indicators.

3.1 Cognitive Modeling of Indoor Environment

A taxonomy proposed by Montello (1993) is introduced to help build cognitive model for indoor wayfinding. The taxonomy contains four levels. Indoor objects

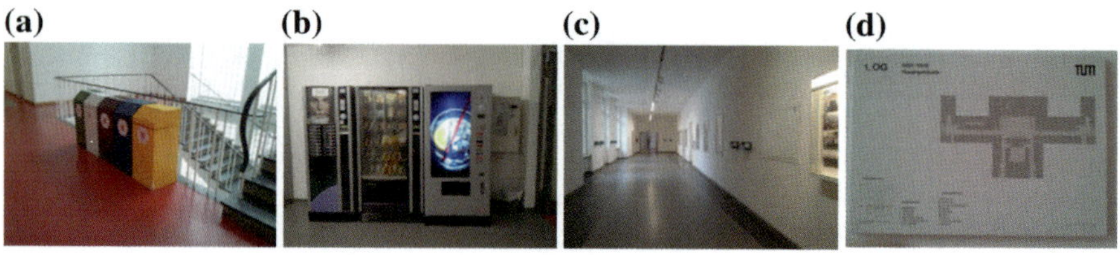

Fig. 1 Examples of indoor object taxonomy, figural objects: **a** a set of movable rubbish bins that are smaller than human body, **b** some unmovable automatic service machines which are slightly larger than the body; vista object: **c** a corridor; **d** environmental object: the floor of a building shown in a vista object—a floor map

can be mapped on three of them, i.e. figural level objects, vista level objects and environmental level objects.

- **Figural** level denotes objects that are smaller than or nearly equal to the size of the human body. People can directly perceive their properties or manipulate them from one place without an obviously location change;
- **Vista** level denotes objects that larger than human body but can be visually perceived in a single view. An example of vista level objects is simple rooms (in this paper we refer to halls and corridors also as rooms);
- **Environmental** level contains objects that are significantly larger than the body. To completely perceive them, people have to make considerable position change. Objects at this level include irregularly shaped rooms, floor configurations and entire buildings.

Figure 1 illustrates some examples about the taxonomy.

The indoor wayfinding is usually performed in two situations: (1) people plan their route across multiple rooms, (2) people want to find a place in a complex room.

In situation 1, people need knowledge to guide their walk through different rooms, thus objects in a single room will get less attention while the relationship between different rooms becomes more important. In this situation, figural and vista objects are abstracted and integrated with the representation of environmental object, i.e. the representation of floor and building. In situation 2, people need knowledge to guide them move around a complex room. Thus figural and vista objects, i.e. the room itself and objects in the room, are more important.

In situation 1, the wayfinding knowledge contains detailed information: routes are defined as potential moving tracks in each room and objects in the room are selected as landmarks. In situation 2, the wayfinding knowledge is more abstract: a route graph is formed by abstracting rooms as nodes and routes as edges connecting the nodes (An illustration is shown in Fig. 2). Landmarks are selected among these rooms.

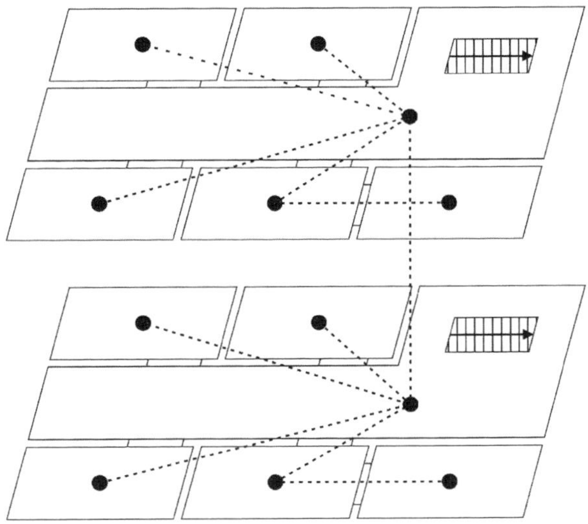

Fig. 2 An illustration of route graph in second situation. *Black dots* denote abstract rooms and *dashed edges* indicate connections between rooms

3.2 A Representation of Indoor Places

The geographical concept of places refers to the areal context of events, objects, and actions. We adopted the definition of affordance-based place (Jordan et al. 1998) and use sextuple to describe indoor objects:

- **Physical features**: appearance, location, configuration of material object(s) can be included in physical features. People perceive affordances from these features;
- **Actions**: affordances that people are possible to perform with objects;
- **Narrative**: historical events or experience;
- **Symbols/names**: symbols or names that denote a specific places;
- **Socioeconomic and cultural factors**: important semantic factors related to people's cultural background;
- **Typology/Categorizations**: places that provide similar affordance for specific people and tasks can be categorized as the same type.

Actions are deduced from a collection of image schemata from certain views by using affordance theory (Gibson 2013). In this process, both physical features and mental interpretation from past knowledge and experiences should be considered (Norman 1988). Raubal and Worboys (1999) depicted a method of extracting affordance from image schemata with the consideration of inaccuracy. For generality, we assume a perfect perception as the knowledge extraction is accurate and complete. The process is shown in Fig. 3: an indoor place is composed of a single object (for example a door makes a place door area) or a collection of objects (for example several benches make a place probably a waiting area); when people perceive a

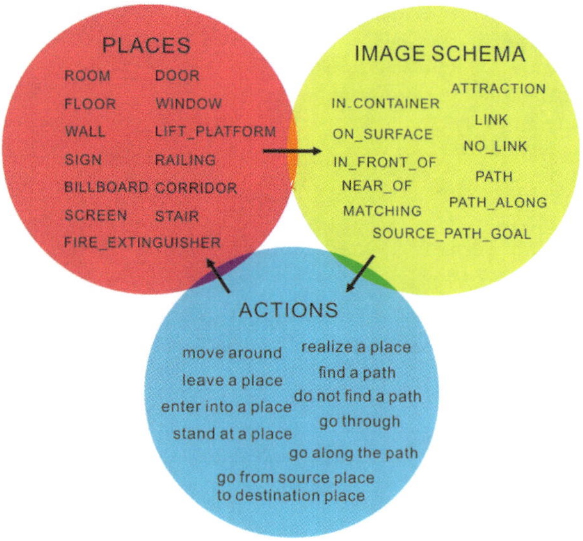

Fig. 3 Relationships between places, image schemata and actions, *arrows* indicate knowledge perceived from physical objects, transformed to cognition patterns and in turn, the cognitive patterns guide people perform actions with objects

certain scene, their observation matches a collection of image schemata. From image schemata people deduce affordances.

Figure 4 illustrates a fragment of how people find their way: (a) we suppose someone is searching for the office of Cartography Department in campus building. He stands in a corridor near his destination, Library ('Bibliothek', room number '1767') of Cartography Department, perceives image schemata IN_CONTAINER (I, corridor) and ON_SURFACE (I, floor), extracts action 'move around'. Then he moves around and is attracted by a billboard and move towards the billboard. The image schemata is ATTRACTION (I, billboard). We assume he immediately realizes the destination room is near here and the sign ensures it from MATCHING (sign, 'Cartography Department'). Then he sees the door and gets LINK (corridor, another corridor), PATH (corridor, another corridor) and action 'go through'. (b) After he goes through the door, he enters into another corridor. As he goes around and searches door by door until he sees the sign by MATCHING (sign, 'destination office').

3.3 Space Syntax and Salience Analysis

Space syntax was conceived by Hillier and Hanson (1984) as a theory to help architects and city planners analyze space configuration. It contains several concepts: *isovist, axial space, convex space* and a collection of analysis tools: *integration, choice* and *depth distance*.

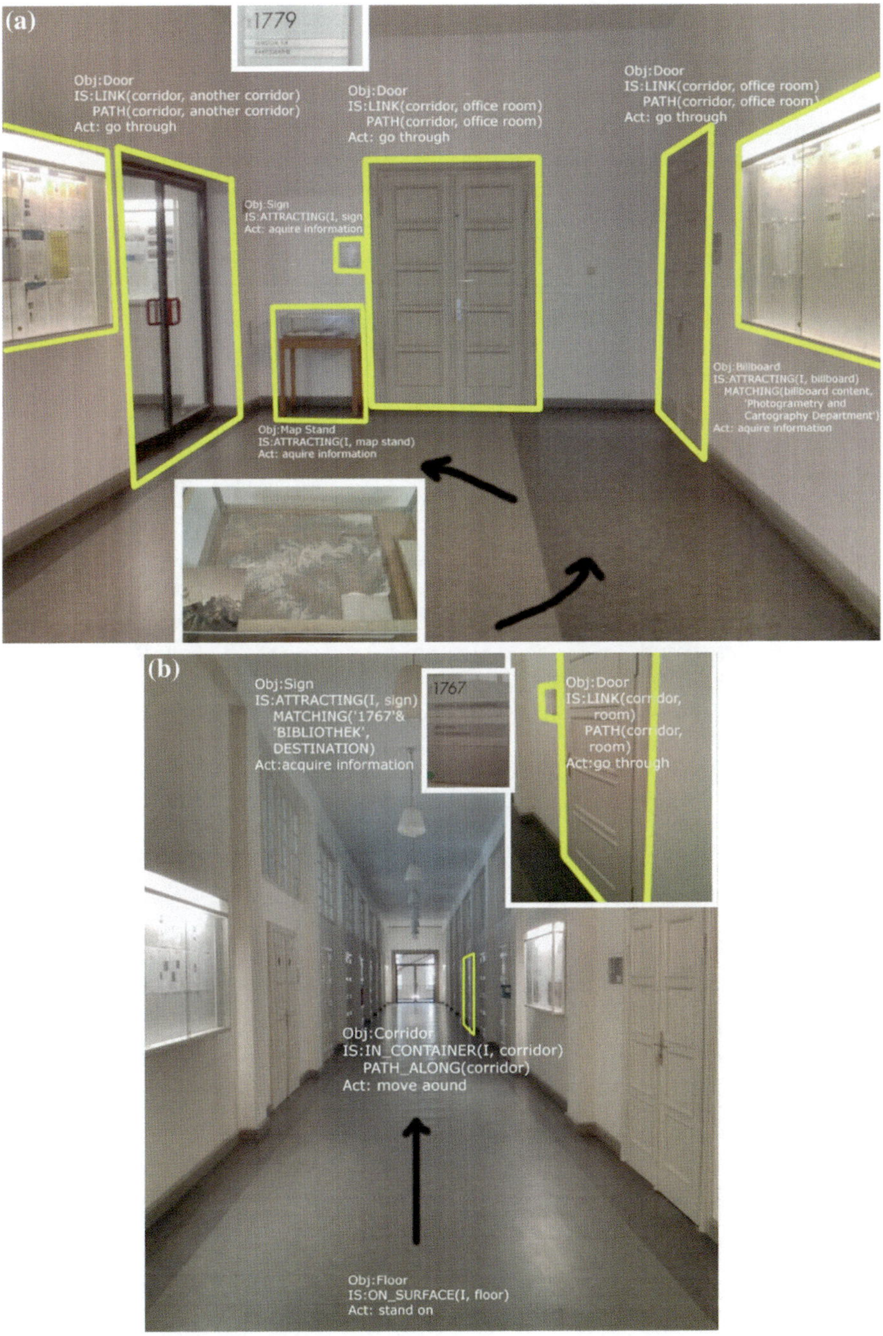

Fig. 4 An example of wayfinding process

The isovist analysis is a popular tool used to identify salient area in buildings. Benedikt (1979) proposed a formal definition with six measures:

(a) area of the isovist,—the area of isovist polygon;
(b) real-surface perimeter of the isovist, indicates how much real-surface can be seen from a vantage point;
(c) occlusivity of the isovist, measures the length of the occluding radial boundary;
(d) variance of the radials is defined as the second moment about the mean of radials' length, it measures the dispersion of the radials length;
(e) skewness of the radials is defined as the third moment about the mean of radials' length, it measures the asymmetry of the dispersion of the radials length;
(f) circularity of the isovist is defined as the isoperimetric quotient of the isovist polygon.

Here, radials are the line segments that 'radiate' from vantage point x to the boundary. The obstacles that prevent the vision from passing through are called real-surfaces. If the measures of all points in the environment are considered as space-varying quantities, a scalar field or 'isovist fields' can be created (An example is shown in Fig. 5). Dalton et al. (2013) discovered public displays at places with large visible area and low jaggedness have maximum salience. Jaggedness is reciprocal of circularity without the constant parameter. Due to the similarity of salience evaluation task, we assume the proposed factors are also applicable to indoor landmarks.

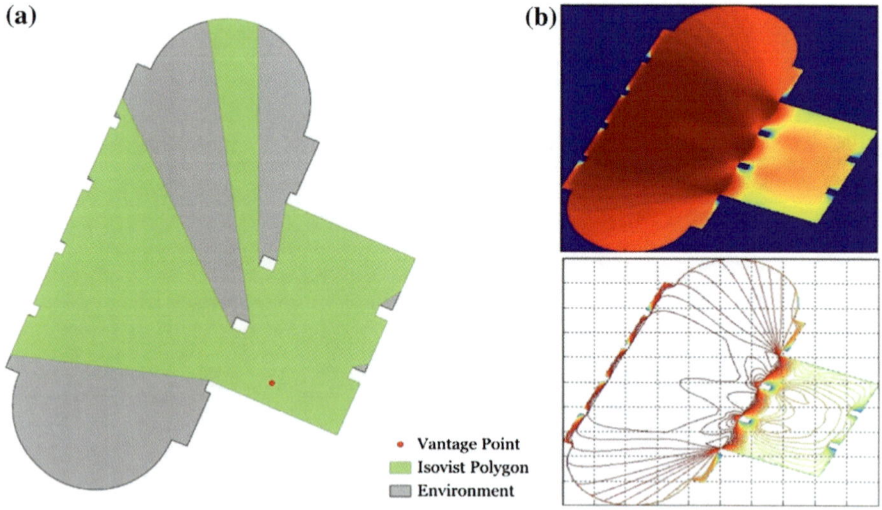

Fig. 5 An example of isovist polygon from a generating location (**a**) and isovist field based on isovist area measure (visualized by a heat map and a contour line map) (**b**)

With regard to the importance of a single room in space configuration, we believe that *integration* and *choice* are the proper indicators, while the linear distance between two room center points does not make much sense.

Based on our description of indoor environment, we adjust a little the definition of these tools:

Integration measures how many rooms have to go through from a starting room to all other rooms in route network as we discussed in Sect. 3.1, using shortest paths;

Choice is simplified as how many other rooms are connected with current room.

4 Salience Indicators for Indoor Landmark

Cognitively prominent is the main character of a landmark. Following former researches, we address the salience from visual, semantic and structural attractiveness.

4.1 Visual Salience

An object is visually salient because it attracts more visual attention than other objects around it. Normally visual properties include shape, color and texture, etc. but there are some specialities for indoor objects.

For figural objects we choose *Visible Area* and *Circularity*, which consists of ***Visual Accessibility***, to measure visual salience as stated in Sect. 3.3. If we use V(x, y) to denote an isovist generated from point (x, y), A(V(x, y)) represents area of V(x, y), and ∂V(x, y) represents the boundary length of V(x, y), the circularity is (as Eq. 1):

$$N_{(x,y)} = \left|\partial V_{(x,y)}\right|^2 / 4\pi A\left(V_{(x,y)}\right) \tag{1}$$

For vista and environmental objects, i.e. rooms, corridors, halls, etc., people usually have difficulties to recognize their orthodox shapes due to their complex shapes. They perceive a collection of isovist polygons as they move from one point to another. We use **Shape Perceivability** to indicate the visual cognition patterns for rooms. *Average Height* and *Variance* can describe the patterns for each isovist field. Formalized definitions for Average Height and Variance are:

We use F denotes a isovist measure field in a region D, F(x, y) indicates the measure value at point(x, y) in D, variance is denoted by Var(F) as (Eq. 2), the average height is denoted by AH(F) as (Eq. 3):

$$AH(F) = \iint_D F(x,y) \cdot dxdy / \iint_D 1 \cdot dxdy \tag{2}$$

$$Var(F) = \iint_D (F(x,y) - AH(F))^2 \cdot dxdy / \iint_D 1 \cdot dxdy \tag{3}$$

Other properties include color, texture and illumination status. Texture is hard to identify and formalize. Illumination status can largely effect visual perception and color perception. Although illumination status can be formally modeled, for example, by counting the IL luminance of a room, it is affected both by the natural light and the indoor lighting system. An ideal solution may be using sensors to capture illumination and color information in real time and convert it into a perception-based color model such as HSV (Nothegger et al. 2004a, b).

4.2 Structural Salience

Accessibility of a room represents the possibility to enter it from as many rooms as possible. As stated in Sect. 3.3, we use *Choice* to measure it. *Choice* can be calculated by counting '*degree*' of each node in the second level route network.

Location importance indicates attractiveness of a room caused by different locations and represented by integration. Integration can be calculated by using the concept '*Closeness Centrality*' from graph theory in our second level route graph. *Closeness centrality* is defined as the inverse of the sum of graph distances between one node and all other nodes (Eq. 4):

$$C_H(x) = \sum_{y \neq x} \frac{1}{d(y,x)} \tag{4}$$

where $C_H(x)$ is the closeness of a node x, d(y, x) denotes distance between two different nodes x and y. Since no two rooms occupy the same space, the distance between any two nodes won't be 0. If there is no route between two rooms, the distance of these two nodes is set to infinity, and its inverse is 0. Actually the definition of closeness centrality not only represents the integration concept, but also is similar to the centrality concept in geometry.

4.3 Semantic Salience

Functional Importance represents the functional attractiveness of an indoor place. The function of a place can be related to the actions that the place offers. We can specify a unique code for each action. By counting the existences of the code, the most common function and the rarest function can be identified. A place with a rare function tends to attract more attention than one with a common function. Thus a scale indicates functional importance can be built.

Other Semantic Properties incl. cultural and historical importance may be derived from narrative and socioeconomic and cultural factors. Although in daily life we used to refer such properties to entire building, it is also possible to specify such properties for single room or object. For example, an office may be historically famous because someone used to work there; a cafeteria may be important in an office building because staff all like the food their; a painting on the wall or a bust in a hall may be a famous artwork. In this paper, we just assign TRUE or FALSE to indicate whether a place has such kind of importance. But we should declared that an extension can be made by using a predefined scale to evaluate their cultural and historical importance as Raubal and Winter did (Raubal and Winter 2002).

4.4 Finding Indoor Landmarks

To estimate whether an object is landmark, its attributes should be compared with attributes of surrounding objects. When selecting global landmarks, candidates are compared with all other objects in the environment. To extract local landmarks, a neighborhood is defined as a subset consists of objects within a certain metric distance from investigated object. With this consideration, we suggest that landmarks are 'outliers' at least over a fraction of all attributes. Hawkins (1980) defined 'outlier' as 'an observation which deviates so much from other observations as to arouse suspicions that it was generated by a different mechanism'.

Outlier detection method can be divided between statistical parametric methods and non-parametric methods. Statistical parametric methods are based on a known distribution or an estimation of unknown distribution parameters while non-parametric methods usually employ data-mining techniques such as distance-based methods and clustering analysis.

In this paper we introduce distance measure functions for each attribute, and comparison functions which indicate the deviation between the current attribute value and the summarized statistic value. We may also assume that the results of comparison functions follow some kind of distribution (e.g. a normal distribution), so the hypothesis test can be done to find out 'outlier values'. With this method, visual, semantic and structural salience can be calculated individually. With predefined weights for each category, each salience measurement can be brought together (Eq. 5):

$$S_{overall} = w_{vis} \times S_{vis} + w_{sem} \times S_{sem} + w_{str} \times S_{str} \quad (5)$$

The predefined weights allow for an adaptation to different scenes.

4.5 *Indoor Landmarks in TUM Inner City Campus Main Building—A Case Study*

In this section we show a case study of TUM inner city campus main building. In practice, it may not applicable to include every object in navigation database. Our dataset contains rooms, Point of Interests (PoIs) and a routing graph. Rooms are captured by their geometry and attributes mentioned in Sect. 3.2. For wayfinding tasks, rooms usually carry affordance related to some movement action, but sometimes a special room type may provide information to help people find their location or orientation. As to objects in a room, only some are captured and stored as PoIs. In Wikipedia PoI is defined as a specific location that interest people, however, the definition is ambiguous. We consider PoI identification is a task-oriented process, the better the affordances fit current task, the more chance it has to become a PoI. For navigation task, PoIs act as locators and (re-)orientation helpers. Locators are not only explicitly signs or public views, but also some furniture, installations or special rooms with which people reason locations by relating background knowledge. Frankenstein et al. (2012); (re-)orientation helpers enable people to get (re-)orientation information by perceiving information from them. Routing graph indicates connections between different rooms as we described in Sect. 3.1.

To calculate each salience measure, a significance score based on z-score (as Eq. 6) is calculated. In this equation x is the individual measurement, μ denotes the mean value, and σ represents the standard deviation.

$$z = (x - \mu)/\sigma \tag{6}$$

Figure 6a shows the rooms and PoIs, PoIs are categorized according to their daily life affordances, Fig. 6b illustrates the salience scores of each PoIs according to their visible accessibility. Bigger dot indicates a higher score.

Five highest visual accessibility scores are shown in Table 1. We found most of these PoIs located in the entrance hall with a complex inner configuration.

Figure 7 visualizes a group of scores for each room on visual, semantic, structural salience and overall scores. Darker colors indicate higher scores. We find a pattern that larger public area in a building with more complex inner structure has higher visual salience score. Corridors and central hall consisting of the sketch of a floor configuration thus gain higher structure salience. In this case, the functional importance does not have significantly variation. The reason may be that room functions are more or less uniformly distributed.

Table 2 lists 6 most salience rooms. As we described, the semantic salience score is not significant even in most salient rooms, while visual and structural salience are two main contributions to the overall salience.

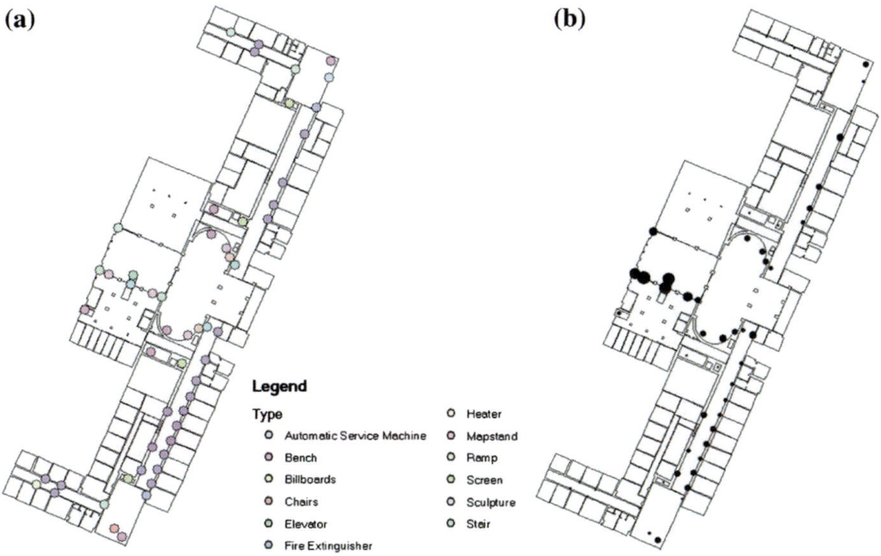

Fig. 6 PoIs on the ground floor of TUM campus main building with type information (**a**) and a visualization of their salience scores with *dots* of different sizes (**b**)

Table 1 Five PoIs with the highest scores of visual accessibility

FID	Type	Sz_area	Sz_nx	Svis
38	Bench	2.71935	3.485369	6.204719
41	Screen	2.812627	1.935486	4.748113
7	Ramp	2.701636	2.028642	4.730278
39	Fire extinguisher	2.85016	1.492741	4.342901
42	Bench	1.761061	1.618805	3.379866

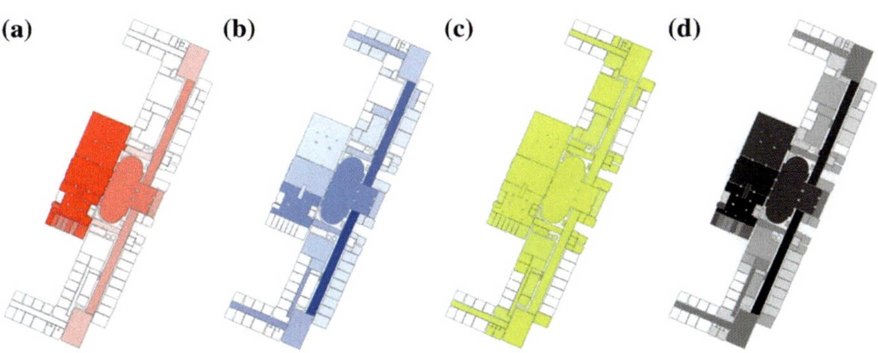

Fig. 7 Landmark extraction scores: **a–c** shows scores for visual, semantic and structural salience; **d** presents overall scores for a combination of the above three salience scores

Table 2 Rooms with the highest overall salience scores and their individual scores on visual, semantic and structural salience

FID	Room function	Svis	Ssem	Sstr	Soverall
70	SSZ INFO HALLE	8.98	0.44	4.62	14.04
23	FLUR SÜD	3.69	0.06	8.95	12.70
29	DURCHGANGSHALLE	9.69	0.25	2.35	12.29
105	FLUR NORD	3.95	0.06	7.15	11.16
30	EINGANGSRAMPE SSZ	9.18	0.38	0.99	10.56
9	EINGANGSHALLE, FOYERS	5.23	0.25	4.53	10.01

5 Conclusion and Future Work

In this paper both theoretical and computational issues related to indoor landmark extraction are studied. Several related theories, such as affordance theory and space syntax, are mentioned to provide salience indicators in this paper. By combining the proposed indicators with predefined weights, we proposed a computational method of extracting landmarks from geo-database. The salience estimation of the case study proves feasibility of our computational model.

The indoor landmark theory is still in its early stage. Progresses on indoor cognition, data collection and modeling may bring some new light to this study. Our study will be deepened in a number of aspects:

1. Cognition of indoor environment: Indoor wayfinding is a cognitive process; the study will serve as a foundation for the salience indicators and for the detection of missing features that affect objects' salience.
2. Multi-sources dataset: Multiple data sources need to be captured and integrated so as to provide more geometric and semantic information than a 2D indoor map. With enriched dataset more features can be captured.
3. Context-awareness: Users may have different sensitivity on the same object according to their preferences, motion types and tasks etc. A context-aware approach concerns such information and selects landmarks individually, thus the usability of a navigation system can be improved.
4. Temporal information: Temporal information includes time (e.g. day or night), weather and other temporal status of indoor objects (for example a corridor with construction work may be temporally blocked). With consideration of such information, regulations under different situation can be modeled and used to provide more practical results.

Acknowledgments The authors gratefully appreciate financial support from China Scholarship Council.

References

Benedikt ML (1979) To take hold of space: isovists and isovist fields. Environ and Plann B 6 (1):47–65

Burnett GE (1998) 'Turn Right at the King's Head': drivers' requirements for route guidance information, Loughborough University

Dalton N, Marshall P Dalton R (2013) Extending architectural theories of space syntax to understand the effect of environment on the salience of situated displays. In: Proceedings of the 2nd ACM international symposium on pervasive displays, ACM

Elias B (2003) Extracting landmarks with data mining methods. In: Spatial Information Theory. Foundations of Geographic Information Science. Springer, Heidelberg, pp 375–389

Frankenstein J, Brüssow S, Ruzzoli F, Hölscher C (2012) The language of landmarks: the role of background knowledge in indoor wayfinding. Cogn Process 13(1):165–170

Gibson JJ (2013) The ecological approach to visual perception. Psychology Press, East sussex

Golledge RG (1999) Human wayfinding and cognitive maps. Wayfinding behavior: cognitive mapping and other spatial processes. JHU Press, Baltimore, pp 5–45

Hölscher C, Meilinger T, Vrachliotis G, Brösamle M, Knauff M (2006) Up the down staircase: wayfinding strategies in multi-level buildings. J Environ Psychol 26(4):284–299

Hawkins DM (1980) Identification of outliers. Springer, Heidelberg

Hillier B, Hanson J (1984) The social logic of space. Cambridge university press, Cambridge

Jordan T, Raubal M, Gartrell B et al (1998) An affordance-based model of place in GIS. Paper presented at the 8th international Symposium on Spatial Data Handling, SDH

Klippel A, Winter S (2005) Structural salience of landmarks for route directions. Springer, Spatial information theory, pp 347–362

Lovelace KL, Hegarty M, Montello DR (1999) Elements of good route directions in familiar and unfamiliar environments. In: Spatial information theory. Cognitive and computational foundations of geographic information science. Springer, Heidelberg, pp 65–82

Lynch K (1960) The image of the city. MIT press, Cambridge

Millonig A, Schechtner K (2007) Developing landmark-based pedestrian-navigation systems. IEEE Trans Intell Transp Syst 8(1):43–49

Montello DR (1993) Scale and multiple psychologies of space. In: Spatial information theory a theoretical basis for gis. Springer, Heidelberg, pp 312–321

Norman D (1988) The design of everyday things/Donald A. Norman Doubleday, New York

Nothegger C, Winter S, Raubal M (2004a) Computation of the salience of features. Spat cogn comput 4:113–136

Nothegger C, Winter S, Raubal M (2004b) Selection of salient features for route directions. Spat cogn comput 4(2):113–136

Presson CC, Montello DR (1988) Points of reference in spatial cognition: stalking the elusive landmark*. Br J Dev Psychol 6(4):378–381

Raubal M, Winter S (2002) Enriching wayfinding instructions with local landmarks. Springer, Heidelberg

Raubal M, Worboys M (1999) A formal model of the process of wayfinding in built environments. In: Spatial information theory. Cognitive and computational foundations of geographic information science, Springer, Heidelberg, pp 381–399

Sorrows ME, Hirtle SC (1999) The nature of landmarks for real and electronic spaces. In: Spatial information theory. Cognitive and computational foundations of geographic information science. Springer, Heidelberg, pp 37–50

Wang RF, Brockmole JR (2003) Simultaneous spatial updating in nested environments. Psychon Bull Rev 10(4):981–986

Part II
Positioning

On the Feasibility of Using Two Mobile Phones and WLAN Signal to Detect Co-Location of Two Users for Epidemic Prediction

Khuong An Nguyen, Zhiyuan Luo and Chris Watkins

Abstract An epidemic may be controlled or predicted if we can monitor the history of physical human contacts. As most people have a smart phone, a contact between two persons can be regarded as a handshake between the two phones. Our task becomes how to detect the moment the two mobile phones are close. In this paper, we investigate the possibility of using the outdoor WLAN signals, provided by public Access Points, for off-line mobile phones collision detection. Our method does not require GPS coverage, or real-time monitoring. We designed an Android app running in the phone's background to periodically collect the outdoor WLAN signals. This data are then analysed to detect the potential contacts. We also discuss several approaches to handle the mobile phone diversity, and the WLAN scanning latency issue. Based on our measurement campaign in the real world, we conclude that it is feasible to detect the co-location of two phones with the WLAN signals only.

Keywords Epidemic tracking · Co-location · WLAN tracking

1 Introduction

In the past decade, mobile phones and the internet have become integrated into daily human life. More importantly, the wireless infrastructure has improved significantly in recent years to help transferring the information amongst the devices.

K.A. Nguyen (✉) · Z. Luo · C. Watkins
Department of Computer Science, Royal Holloway, University of London,
Surrey TW20 0EX, UK
e-mail: khuong@cantab.net; me@khuong.uk

Z. Luo
e-mail: zhiyuan@cs.rhul.ac.uk

C. Watkins
e-mail: chrisw@cs.rhul.ac.uk

G. Gartner and H. Huang (eds.), *Progress in Location-Based Services 2014*,
Lecture Notes in Geoinformation and Cartography,
DOI 10.1007/978-3-319-11879-6_5

People leave real-time digital footprints everywhere. These footprints may be used to track and study a disease in an epidemic. However, this information is largely unexplored in the public health research community so far. Since most people have a mobile phone, a contact between two persons can be regarded as a handshake between the two phones. Our task is to detect when the two mobile phones are close. In this paper, we propose a new methodology to allow the user to passively discover his contacts with other people in an epidemic. Our approach does not require a map or real-time GPS signals.

The key features of our approach are:

- We detect co-location—when two phones are close together without detecting exactly where the two phones are.
- This is done passively by an App on the user's smart phone. No signals are sent out.
- The users have full control of whether to track themselves.
- The tracking process uses existing Wi-Fi Access Points, and we have tested it in challenging, uncontrolled city environments.

The paper first explains how our idea can be applied into the epidemic tracking purpose. Then we discuss several wireless signal candidates for our system, and explain how to collect such signal on an Android phone. The properties and the challenges of using the signal are discussed. Finally, we conclude our findings and outline the future work.

2 Wireless Tracking for Epidemic Detection

2.1 Co-Location Tracking with Mobile Phones

Co-localisation is the process of identifying if two persons are in the same position at the same time. If they are co-located, there is a possibility that one can be infected by the other's disease. Given a time-stamp, we can keep track of the disease spreading history. Imagining a network of registered participants, where each patient uses his mobile phone to input his current symptoms. This information can be later uploaded onto a central server, and the system works out the probability of what disease he was infected. To discover the origin of an unknown disease, the doctor back-tracks the patients' contact history with other registered people in the same system.

Since most people have a mobile phone, a physical contact between two persons can be regarded as a handshake between the two phones. When such contact happens, the handsets must be close to each other. With our idea, we need not maintain a map for the devices, nor require knowing the exact location of the phone at any moment. The remaining question is: 'How can we detect when the mobile phones are close?' Given a particular time-stamp, we need a unique and ubiquitous property to reliably match any two mobile phones' location, and recording that a physical contact has happened. Fortunately, there are many wireless signals such as WLAN, Bluetooth, GSM, FM that are available in many places and can be freely captured. The next section discusses the pros and cons of these wireless signals for our project.

2.2 The Wireless Signals Candidates

Signal availability is the most important criterion to decide which wireless signal we will use. The signal should cover both indoor and outdoor spaces. Two stand-out candidates were the WLAN signal, and the GSM signal.

The GSM signal coverage has increased significantly in recent years, thanks to the wide deployment of many cellular phone towers (Ibrahim and Youssef 2013). Unfortunately, the Android NeighboringCellInfo class used to access GSM cellular information is phone-dependent and network-dependent. Many Samsung phones did not work in our experiments.

The WLAN signals are popular indoor, but were not so popular outdoor a few years back. Thanks to the increasing number of outdoor WLAN Access Points (APs), most notably the recent BT-Fon network, which allows the home router to transform into a public wireless hotspot. There have been over 5 million available APs in the UK since 2012, with 20,000 new hubs being added weekly (represented by the red and blue dots in Fig. 1). In our experiments, there are always at least 10 available APs at any position. Many areas in the city centre of London have more

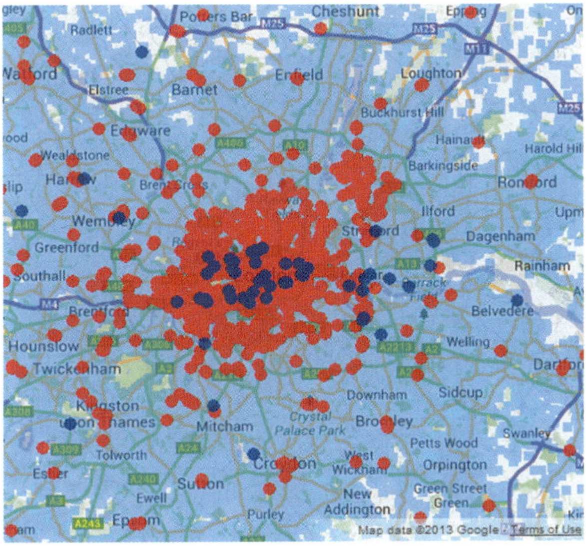

Fig. 1 BT WiFi hotspots in London

than 30 accessible APs. Another example is the commercialised Skyhook project,[1] which provides a world-wide WiFi RSSI signal strength to physical location map to alleviate the need of GPS coverage.

Some popular signals such as Bluetooth or infrared have restricted range, and are not popular outdoors. Other signals such as FM are available, but require additional decoder on the handset to read them. Such decoders are not widely supported by current smart phones. Thus, the WLAN signal is our best candidate. It is also possible to combine other signals with WLAN to increase the location's uniqueness (Pei et al. 2012).

2.3 An Android App to Collect the WLAN Signal Strength

To record the WLAN signals, we designed an Android app, which runs in the phone background to periodically scan the signals from the nearby APs every 30 s. This is the default scanning rate, which can be customised. We chose the Received Signal Strength Indicator (RSSI), which can be collected easily with the Android API.

The theoretical WLAN RSSI varies from 0 to −100 dBm, where higher value represents stronger signal. We looked into the Android source code, and found that RSSI equal or bigger than −55 dBm is considered the strongest signal, which is shown as a full bar of signal on the phone. If the measured RSSI is equal to or less than −100 dBm, the Android phone shows an empty bar. Based on this scale, we define three ranges of WLAN RSSI.

- RSSI from −55 to −70 dBm represent strong signals.
- RSSI from −70 to −85 dBm represent medium signals.
- RSSI from −85 to −100 dBm represent weak signals.

Under normal usage, our app consumed 29 % of the total power with our Google Nexus phone (Fig. 2a), and 34 % on our Galaxy Y phone (Fig. 2b).

2.4 How to Collect WLAN Signals on Android: Active Scanning or Passive Scanning?

There are two ways to collect the WLAN signals with an Android phone, active scanning and passive scanning. Both of them belong to the IEEE 802.11 MAC layer. In both cases, there is no authentication needed between the mobile device and the AP.

[1] http://www.skyhookwireless.com.

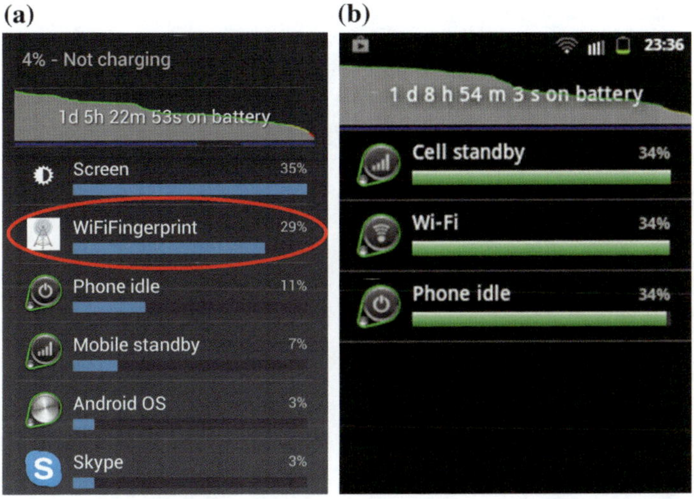

Fig. 2 Battery consumption of WiFi scanning. **a** Galaxy Nexus. **b** Samsung Galaxy Y

With passive scanning, the phone constantly listens on consecutive channels for the beacons periodically sent by the APs. We looked into the Android source code, and found that the dwelling time on each channel is set at 120 ms. According to the IEEE 802.11 standard, the WLAN APs should send out beacons every 100 ms on all channels at the same time. Theoretically, the extra 20 ms interval should be sufficient for the mobile device to receive at least one beacon per channel. Thus, with 13 channels of the 2.4 GHz spectrum (in Europe), it takes at least 1,560 ms for an Android phone to scan all nearby APs theoretically.

In reality, the total scanning time also includes the information processing delay, in which the device processes the received beacons on each channel. Although passive scanning consumes less battery power, the device cannot pick up hidden APs, which are configured not to send out any beacon. Table 1 compares the passive scanning time in 5 h with our two phones.

With active scanning, the mobile device sends the probe request frames on all channels (similar to how the beacons are sent by the APs), and waits for the probe responses from the APs. The probe request frame can either contain a network name (SSID) of the AP the mobile device wishes to connect to, or an empty SSID, which all nearby APs should respond to. According to the IEEE 802.11 standard, the device should listen for a minimum of MinChannelTime (in ms) on a single

Table 1 Summary of passive scanning time

	Google Nexus (ms)	Samsung Galaxy Y (ms)
On average	5,022	5,016
Slowest single scan	5,194	5,641
Fastest single scan	4,909	4,902

Table 2 Summary of active scanning time

	Google Nexus (ms)	Samsung Galaxy Y (ms)
On average	962	1,419
Slowest single scan	1,045	1,532
Fastest single scan	944	1,400

channel. If no probe response is heard within this interval, the device assumes that this channel is empty, and moves on to the next one. If more than one probe response is heard within this interval, the device will continue to listen till the MaxChannelTime (in ms) has elapsed on the same channel. There is no strict definition of MinChannelTime and MaxChannelTime by IEEE 802.11 however. We looked into the Android open source code and found that Google implemented just a single dwelling time constant of 30 ms. For each probe request frame the device sends on the channel, there is an extra 3 ms delay. Table 2 compares the active scanning time in 5 h with our two phones.

In reality, laptops and other devices can decrease the scanning latency by forcing a scan on a specific channel only. Therefore, the device constantly listens on a particular channel, knowing that the AP should send out beacons on all channels. However, we cannot execute this method on non-modified Android firmware yet.

With either scanning setting, our app only wakes the CPU up every 30 s to scan the WLAN signals. This parameter is also customisable. While active scanning can complete faster and discover hidden APs, it consumes more power than passive scanning. A better option is to perform active scanning when the user is on the move, and switch to passive scanning when no movement is detected.

2.5 The Inverted System

Theoretically, it is possible to invert our system to have the APs track the mobile phones. This structure has the advantage of requiring no additional code on the phones, at the expense of a higher processing load on the APs, which have to track a huge amount of mobile users. In addition to the scalability issue, it is unlikely such changes can be done on the APs, without permission from the network providers.

On the security side, it would be undesirable for the users to be forcibly tracked by the APs. The users' anonymity is broken, because he can be identified by his GSM number or the phone's WLAN MAC address. In contrast to our original approach, the users can simply stop the tracking app on the phone without disrupting normal GSM or WiFi uses.

Table 3 Comparison of our system and FluPhone

	Our system	FluPhone
Technology	WLAN	GPS and Bluetooth
Contact detection	Off-line	Real-time
Battery consumption	Average	High
Always-on connection	No	Bluetooth
Custom code	Yes	Yes

2.6 Related Work

To the best of our knowledge, there was only one similar research known as the FluPhone project (Yoneki 2011). Both of our system and FluPhone aimed to provide mobile phone localisation for epidemic tracking. However, there are two key differences. First, we do not record the physical location of the users. Second, while FluPhone uses GPS and Bluetooth signal to discover nearby handsets in real-time, our approach analyses the off-line signal data to discover such contact. Table 3 compares the two approaches.

There were other work involved the WLAN signals for indoor and outdoor localisation (Wang et al. 2012; Chintalapudi et al. 2010; Martin et al. 2010). Yet, there was little attention to the co-localisation aspect, where the timing of the contact is important (Krumm and Hinckley 2004). The most notable use of WLAN localisation was fingerprinting, where real-time WLAN data are compared to a training database (Bahl and Padmanabhan 2000). Our work does not involve such training data, instead, the signal from different phones are compared directly based on its time-stamp.

3 Test Beds

We recorded the WLAN signals in two UK cities. Our first test bed was recorded near a busy railway station, where the second test bed was recorded on the streets of London. We used two Android mobile phones—the Google Nexus with the Jelly Bean firmware, and the Samsung Galaxy Y with the Gingerbread firmware. The 'WiFi Fingerprinting' app we designed to collect the WLAN signals can be downloaded on the Android app store.

4 WLAN Signal Properties

With our app, we collected the WLAN signal in different locations in the UK to assess the following criteria for both static and moving phones.

- Two co-located phones should observe similar WLAN signals.
- How the WLAN signals distinguish in different locations.

While other research investigated the indoor WLAN properties for fixed clients, our experiments looked at the signal properties for outdoor moving device, where the environment is not as stable as indoor. Further, we are more interested in how distinguishable multiple signal traits are, rather than from a single device perspective as in other work. Finally, we focus on the number of found APs, beside the signal strength. We used the RSSI as a quality measurement for the WLAN signal in our experiments.

4.1 The WLAN Signal of Static Phones

When the two phones are co-located, we assume that they should hear the same signals from nearby APs. Figure 3 depicts the histogram distribution of the WLAN signals between our phones, which are positioned right next to each other, and an outdoor BT AP. We collected 6,130 signal readings in half a day. In our experiment, the signal variation was around 20 dBm, in contrary to the 10 dBm interval reported for indoor WLAN (Kaemarungsi and Krishnamurthy 2004). The maximum and minimum readings observed from Google Nexus phone were −60 and −77 dBm respectively, while Galaxy Y recorded −56 and −75 dBm. However, the majority of signals peak around the highest frequency RSSI in both phones. The most frequent RSSI was −64 dBm for Google Nexus, and −63 dBm for Samsung Galaxy Y.

Despite the total 20 dBm signal variation, the maximum signal difference at any moment between the two phones was under 11 dBm (Fig. 4). The signal difference was small at night, and became bigger by lunch time.

Figure 5 shows that the signal difference of the two phones was less than 4 dBm for 91 % of the time. They had the exact signal reading for 16 % of the time. The majority of the signal difference was just 1 dBm, 34 % of the time.

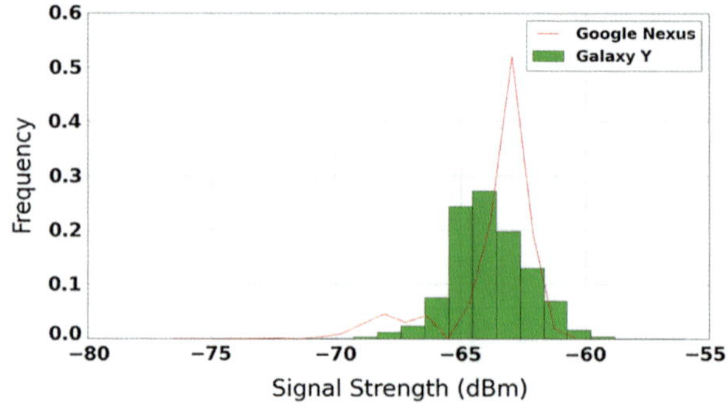

Fig. 3 Outdoor WLAN signals distribution

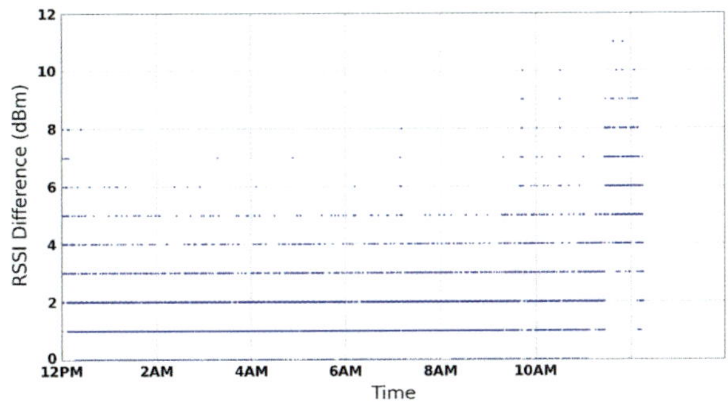

Fig. 4 Individual signal strength reading between two phones

4.2 *The WLAN Signal of Moving Phones*

As the user moves around, the scanning latency affects the similarity of the receiving signal in different locations. The latency is caused by the delay from the handset in sending the probe request frames, and the delay from the APs to reply with response frames. For example, AP_1 responses within the first 30 s, however, AP_2 and AP_3 response a second later, when the user has already moved to a new position.

In our experiment, two persons walked side by side with a mobile phone in the pocket. Both phones were synchronised to invoke 3 continuous scans every 30 s. Our Google Nexus phone took less than 1 s on average per scan, while the Samsung Galaxy Y phone took more than 1.4 s. The continuous scans help discovering the missing APs from previous scans. Figure 6 depicts the WLAN signal strength from the phones to the nearest fixed AP. Since we did not know the exact distance from the AP to our handset, we assumed that the starting point is the location where the strongest signal can be obtained, indicated by the zero point on the x-axis.

We observed that the variation is large for stronger signals, and decreases as the signals fade away. The signal was completely lost among all the ambient noises at a distance of 170 m, from the initial position. This result shows a much greater range with outdoor AP, compared to the relatively short 15–20 m distance for an indoor AP.

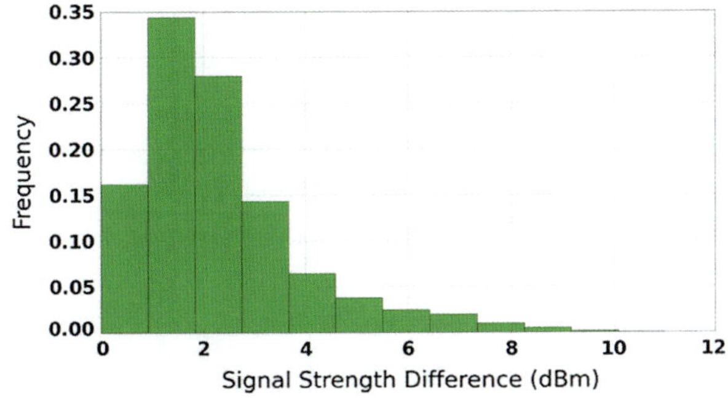

Fig. 5 Percentage of signal reading difference between two phones

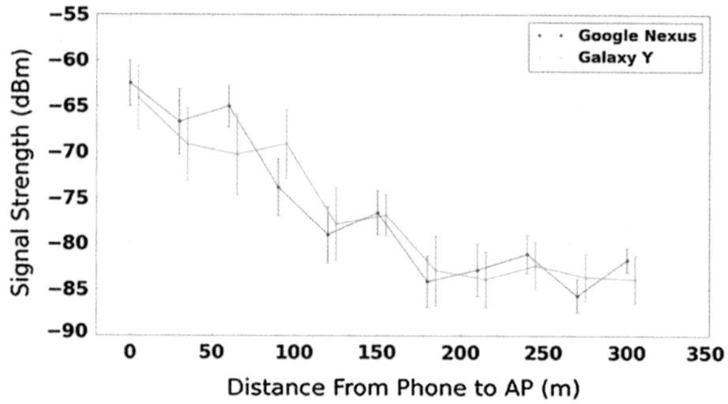

Fig. 6 Correlation between WLAN signal strength and distance

We expect a rough 5–10 dBm difference between two consecutive locations 30 m apart, assuming they are in a relative straight line from the AP.

Since both phones were side by side in our experiment, we expect them to discover the same number of APs. However, *none of the 253 pairs of signal vectors between the two phones has the same number of detected APs*. This number highly contrasted with 3,107 pairs of WLAN signal vector with the exact number of detected APs, over 3,894 total pairs, recorded when the two phones are not moving in the previous experiment. This result confirms the scanning latency we suspected above.

4.3 Number of Detected APs

One of the indicators to distinguish the phone's location is the amount of detected APs. In our experiment, there were 412 APs recorded by the Google Nexus phone, and 554 APs recorded by the Samsung Galaxy Y phone. Of 252 APs found by both phones during the 20 min journey, the highest number of appearance of a single AP was 15 for Google Nexus and 13 for Samsung Galaxy Y (Table 4).

Figure 7 demonstrates the appearance of the top 10 commonly observed APs in the whole journey of both phones. Those APs appeared frequently in both phones' list of detected APs, when they are in the same position (started at 14:56). The Google Nexus phone has a stronger and newer antenna, therefore, it can discover more APs than the old Samsung Galaxy Y phone.

Table 4 Number of detected APs

	Google Nexus	Samsung Galaxy Y
Total number of scans	120	120
Total recorded APs	504	558
Commonly observed APs	252	252
Highest number of single AP found	15	13

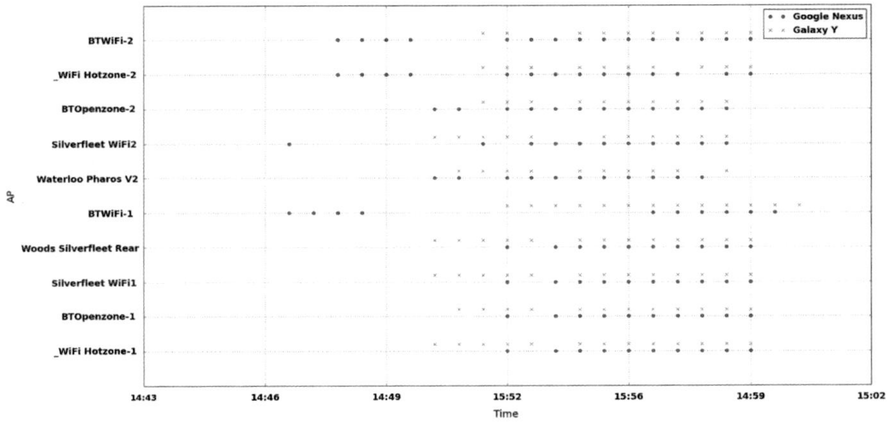

Fig. 7 Top 10 AP with highest frequency of appearance

5 Mobile Phone Co-Localisation with WLAN Signals

5.1 A Matching Rate Algorithm for Co-Located Phones

In our experiment, we recorded 1,643 APs in 1 h journey in London. We observed at least 10 APs at any given place, and more than 30 APs in the city centre of London. Based on such high number of APs, we define an algorithm to calculate a matching rate given any two RSSI vectors.

The two WLAN vectors are considered 100 % matched, if they have the same number of recorded APs, and for every AP recorded by Phone_1, it is also observed by Phone_2, and vice versa. If the two signal vectors share some APs in common, but also contain their detected APs. The matching rate is calculated as the number of common APs, divided by the total number of APs. A high matching rate means the two devices are close, and a low one means they are further away. Ideally, we aim to deliver a matching rate as close as possible to the physical distance.

The above algorithm worked relatively well for a large number of APs. When the number of nearby APs is low, we try to incorporate the WLAN signal strength into our equation. Our assumption is that when two phones are close, they should observe a strong signal from the same AP. In other words, if one phone sees a strong AP, while the other does not, they are unlikely to be close. However, this assumption is subjective, and although it works well in our experiment, certain location with different combinations of APs may not see a better result. We add up the signal strength of the common APs from the WLAN vector of both phones. The result is divided by the total signal strengths from all nearby APs of both devices.

In our experiment, two persons started at the same location, they then walked in different paths at 14:42, and re-joined midway at 14:56, and continued the journey until 15:01 (Fig. 8).

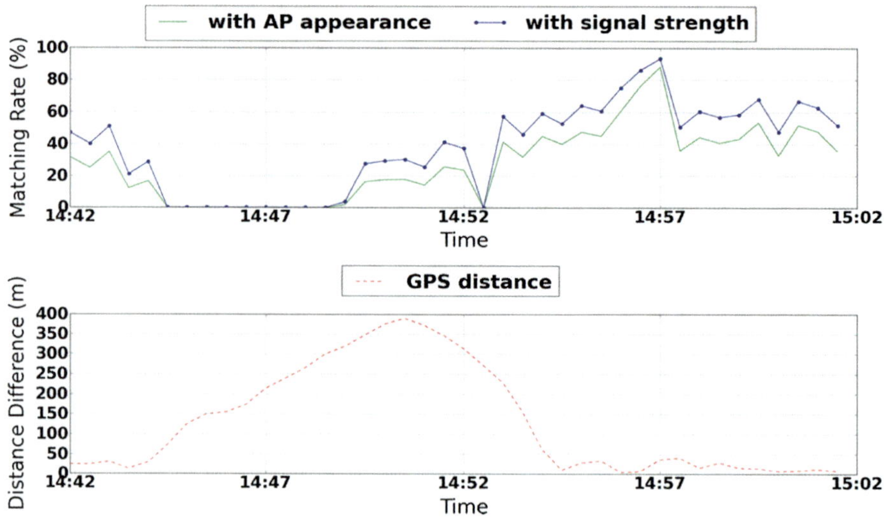

Fig. 8 Comparison of matching rate and GPS distance

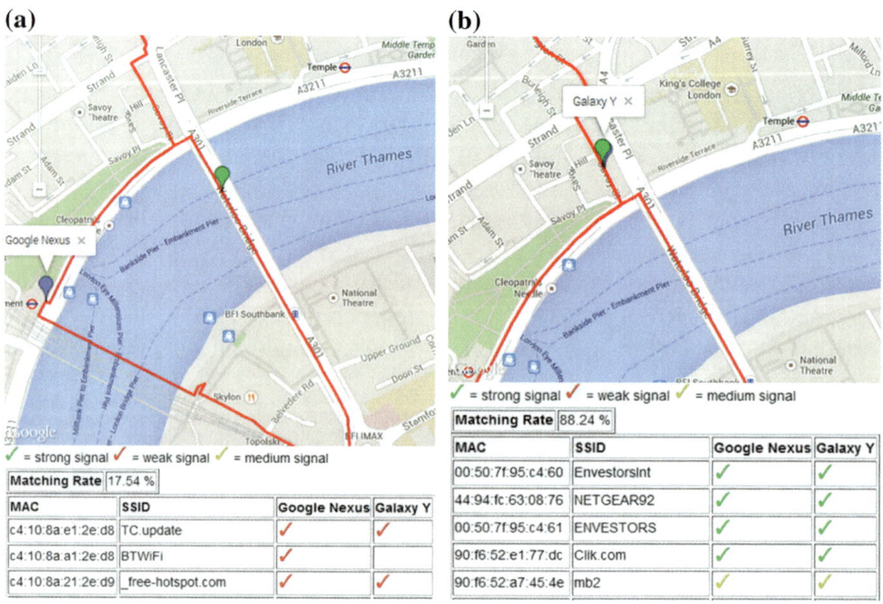

Fig. 9 Mobile phones co-localisation on google maps. **a** Low matching rate for separated phones. **b** High matching rate for co-located phones

The matching rate started off positively at the beginning, when both phones were together. As the phones went in different routes, the matching rate dropped. Since both phones were in open space, they still saw some similar APs, despite their distance (Fig. 9a). However, the observed RSSI was weak. The matching rate

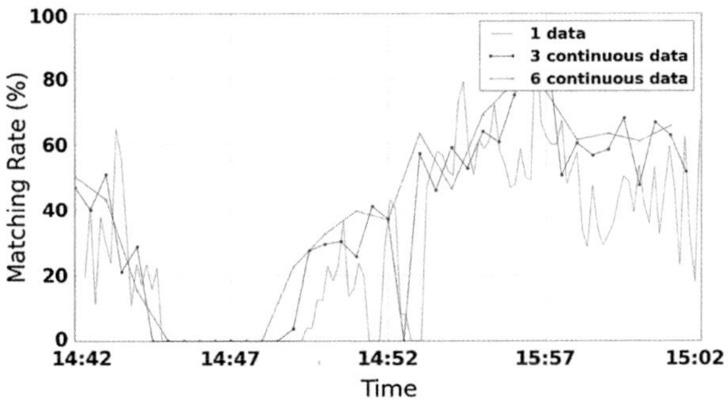

Fig. 10 Improvements from multiple continuous scans

increased as the phones approached each other, and remained stable for the remaining journey (Fig. 9b). When the phones are close, they both observe many strong RSSI. A real-time demo of our experiment on Google Maps can be viewed on our website.[2]

5.2 Handling the Mobile Phone Signal Diversity

Each manufacturer can implement their WLAN antenna, or WLAN adapter, which can potentially affect the receiving signal on the mobile device. A new mobile phone with a bigger and stronger antenna is able to discover long-distance APs, and receives stronger signals from nearby APs. There were multiple attempts to tackle the heterogeneous devices issue, which we classify into two broad categories, calibration-based and algorithm-based. In the first type, the new device is calibrated either manually or automatically within the system. Lee and Han (2012) use different known landmarks in the building to calibrate the devices' signal. With algorithm-based approach, Park et al. (2011) use a linear transformation model and kernel estimation were used to solve the problem, while Kjærgaard (2011) and Ibrahim and Youssef (2013) compare two pairwise signal vectors directly. In this paper, we explain our simple approach to normalise the WLAN signal data, since our goal is off-line co-localisation detection and we already have access to the full database through-out the phones' journey.

Our approach assumes that all registered devices may observe the strongest possible signal within their antenna capability. This is a fair assumption, given the vast number of APs, and the amount of the time the phone is observed. We normalise the signal strength readings for each mobile phone into a number between (0, 1), so that they are directly comparable to the signal strengths from other devices. Without loss of generality, given an RSSI vector representing the WLAN signal strength of N nearby APs, $RSSI = (s_1, s_2, \ldots, s_N)$, with s_i is the signal strength observed from AP_i, we divide each s_i by the strongest signal $s_{\max}$ observed in the whole journey.

[2] http://khuong.vn/Map.

$$
\begin{aligned}
RSSI_{norm} &= \left(\frac{s_1}{s_{\max}}, \frac{s_2}{s_{\max}}, \ldots, \frac{s_N}{s_{\max}}\right) \\
&= (s'_1, s'_2, \ldots, s'_N)
\end{aligned}
$$

5.3 Handling the Scanning Latency of Moving Phones

The movement and speed of the user have a strong impact on the AP discovery and the signal strength. We have shown that even when the two moving phones were side by side, none of our recorded signal traits was 100 % matched at any moment, based on the appearance of the APs. However, 79.8 % of the pairs had 100 % matching rate, when the phones were not moving. A simple approach is to increase the time window, and the continuous scan frequency to capture the missing APs. Since the user cannot move a long distance in a short period of time, and the WLAN signal strength was similar within 30 m in our experiment, we can combine multiple scans within 10–30 s, depending on the walking speed. We outline our scheme to combine *N* continuous WLAN scans.

- First, we generate a list of all APs found within N continuous scans.
- Second, we average the signal of the same APs found within these scans.
- Finally, we remove the APs with very weak signal (e.g. ≤ -90 dBm).

Since a single active WLAN scan takes 900–1,500 ms in our experiment, the parameter *N* should not be too big to conserve battery power, but is also not too small to capture the missing APs. We decided to leave out the weak APs, because some old phones may not be able to see them, as with our Samsung Galaxy Y.

With our method, the number of perfect matching pair of WLAN vectors increased from 0 to 5. The number of 50 % matching pair of vectors increased from 46 to 114. We increased the majority of the matching rate to 40 % with our approach. Figure 10 shows a higher matching rate after combining 3 and 6 continuous scans.

5.4 Bringing It All Together

Figure 11 demonstrates the progress of our scheme for co-localisation detection, given the WLAN signal database from two phones. First, we normalise the WLAN signal strength with our device diversity handling scheme (Sect. 5.2). Several continuous scans are then combined to tackle the scanning latency (Sect. 5.3). In the processing phase, a matching rate is calculated, for each pair of WLAN vectors with the same time-stamp (Sect. 5.1).

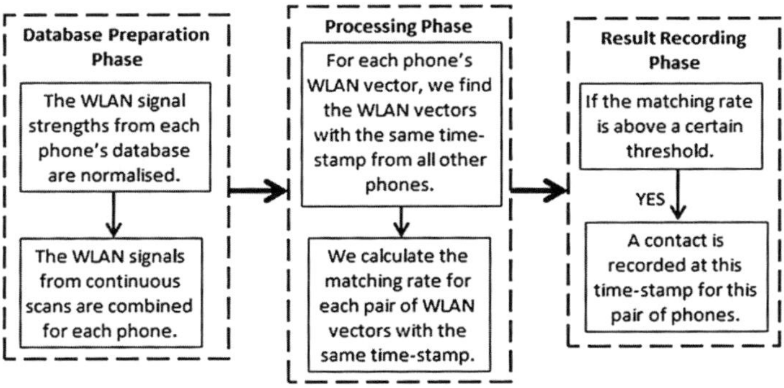

Fig. 11 The progress of our co-localisation scheme

6 Conclusion

We have demonstrated the feasibility of co-localisation detection using mobile phones and the public outdoor WLAN APs for the epidemic tracking purpose. We designed an Android app to collect the WLAN signals, and investigated their properties for co-localisation tracking. To evaluate our approach, we defined a method to compute matching rate, and compared it to the actual GPS distance. We tested our approach in a real, crowded environment to confirm that the matching rate closely reflects the GPS distance. We discussed our approach in handling the mobile device diversity by normalising the WLAN signals. We also identified the scanning latency which is the main cause of degrading the matching rate of moving phones. We showed that the matching rate can be improved up to 30 % by combining multiple continuous scans within a small time-window. Our future work is to investigate new ways of computing the matching rate even with a small number of Aps, and conduct a large scale experiment with many phones.

Acknowledgments The authors would like to thank the anonymous reviewers for their insightful comments on the paper. This research is funded by the Computer Science Department of Royal Holloway, University of London, and EPSRC grant EP/K033344/1 ("Mining the Network Behaviour of Bots").

References

Bahl P, Padmanabhan VN (2000) RADAR: an in-building RF-based user location and tracking system. In: Proceedings of nineteenth annual joint conference of the IEEE computer and communications societies, INFOCOM 2000, vol 2. IEEE, pp 775–784

Chintalapudi K, Padmanabha Iyer A, Padmanabhan VN (2010) Indoor localization without the pain. In: Proceedings of the sixteenth annual international conference on mobile computing and networking ACM, pp 173–184

Ibrahim M, Youssef M (2013). Enabling wide deployment of GSM localization over heterogeneous phones. In: IEEE international conference on communications, pp 6396–6400
Kaemarungsi K, Krishnamurthy P (2004) Properties of indoor received signal strength for wlan location fingerprinting, in mobile and ubiquitous systems: networking and services. The first annual international conference on obiquitous. IEEE, pp 14–23
Kjærgaard MB (2011) Indoor location fingerprinting with heterogeneous clients. Pervasive Mob Comput 7(1):31–43
Krumm J, Hinckley K (2004). The nearme wireless proximity server. In UbiComp: ubiquitous computing. Springer, Heidelberg pp 283–300
Lee M, Han D (2012) QRLoc: user-involved calibration using quick response codes for Wi-Fi based indoor localization. In: 7th international conference on computing and convergence technology (ICCCT). IEEE, pp 1460–1465
Martin E, Vinyals O, Friedland G, Bajcsy R (2010) Precise indoor localization using smart phones. In: Proceedings of the international conference on multimedia, pp 787–790
Park JG, Curtis D, Teller S, Ledlie J (2011) Implications of device diversity for organic localization. In: Proceedings of INFOCOM. IEEE, pp 3182–3190
Pei L, Liu J, Guinness R, Chen Y, Kroger T, Chen R, Chen L (2012). The evaluation of WiFi positioning in a bluetooth and WiFi coexistence environment. In: ubiquitous positioning, indoor navigation, and location based service (UPINLBS). IEEE, pp 1–6
Wang H, Sen S, Elgohary A, Farid M, Youssef M, Choudhury R (2012) No need to war-drive: unsupervised indoor localization. In: Proceedings of the 10th international conference on mobile systems, applications, and services, ACM, pp 197–210
Yoneki E (2011) Fluphone study: virtual disease spread using haggle In: Proceedings of the 6th ACM workshop on challenged networks, ACM, pp 65–66

3D Indoor Location on Mobile Phones Using Embedded Sensors and Close-Range Photogrammetry

Xiujuan Li, Yan Zhou and Hanjiang Xiong

Abstract Indoor positioning on mobile phones has become more and more important to many applications. This paper presents a sensor-based 3D positioning system on mobile phones. It also proposes an efficient approach to determine the initial position which has a great influence on the position precision. By taking a photo of the pre-deployed cross board, the camera's position can be identified immediately using the principle of single-photo resection. The proposed positioning scheme performs location estimation in three phases. First, use cross boards to precisely determine the initial position. Second, employ accelerometer, magnetometer embedded in mobile phones and INS to track the user's position automatically. Finally, visualize the user's location in the 3D model of the building based on GPU rendering technology. The experiment carried out in this paper has good results and indicates that the new method with the embedded sensors and close-range photogrammetry is a promising solution for indoor location.

Keywords 3D indoor visualization · Mobile devices · 3D indoor location · LBS · Close-range photogrammetry · Inertial navigation · Sensors

1 Introduction

With an increasing demand for location services and the development of indoor modeling, 3D indoor positioning has received great attention from both research and industry. The widespread use of smart phones deluges them with demand for

X. Li · Y. Zhou · H. Xiong (✉)
State Key Laboratory of Information Engineering in Surveying, Mapping and Remote Sensing, Wuhan University, 430072 Wuhan, China
e-mail: Xionghanjiang@163.com

X. Li
e-mail: 136596387@qq.com

Y. Zhou
e-mail: zhouyan19910308@gmail.com

G. Gartner and H. Huang (eds.), *Progress in Location-Based Services 2014*,
Lecture Notes in Geoinformation and Cartography,
DOI 10.1007/978-3-319-11879-6_6

location based services inside buildings, which in turn requires mass indoor localization technologies based on mobile phones.

Many infrastructure-dependent technologies such as WiFi (Prasithsangaree et al. 2002), RFID (Bahl and Padmanabhan 2000), ultrasound (González et al. 2009) have been implemented to provide indoor positioning services. However, these methods rely on huge pre-investment to equip the devices and therefore could not serve as a widespread solution. GPS plays an important role in outdoor location service (Bajaj et al. 2002), but its signal is often blocked inside the buildings. Therefore it is limited when the users are indoors at the beginning.

The objective of this work is to develop an accurate indoor positioning system by employing modern mobile phones' embedded sensors like accelerometer, compass and camera, which can run on mobile phones independently without the requirement of additional hardware or external infrastructure. The principle of this indoor positioning system is based on inertial theorem, which is calculated via the data collected by accelerometer and gyroscope. Furthermore, a calibrating method is proposed to precisely determine the initial value and to correct accumulated error caused by sensors. The method uses the principle of Single Photo Resection in photogrammetry to obtain the user's current position.

This paper contributes to the fields of 3D indoor positioning by presenting a system which provides an interactive and friendly mobile user experience. The outline of the paper is as follows: Sect. 2 presents briefly related works about GPU based 3D indoor visualization on mobile phones, Sect. 3 describes a new method of determining initial position precisely using close-range photogrammetry and the way to provide continuous location service based on sensors, Sect. 4 shows the results of our experiment and evaluates its efficiency.

2 3D Indoor Visualization on Mobile Phones

The system aims to provide an interactive and friendly mobile user experience based on 3D model visualization technique, which helps in better understanding spatial relations like where to find the desired merchandise in a shopping plaza, how to reach the specified gate in the airport, which parking space is available in the parking lot. Concerning the pervasive, portable and affordable characteristics of mobile phones and the inevitable development of indoor 3D positioning technology, an experimental system that visualizes the 3D indoor models and user's walking route on android based mobile phones has been developed. The system is implemented through Android JNI by using C++ language and OpenGL ES library.

The 3D scene consists of three levels as shown in Fig. 1.

1. Global Scene: the visualization of the global image data organized with quad-tree pyramid structure.
2. AOI Scene: the target building model which you'd like to enter.
3. Floor Scene: the specified floor model in the building.

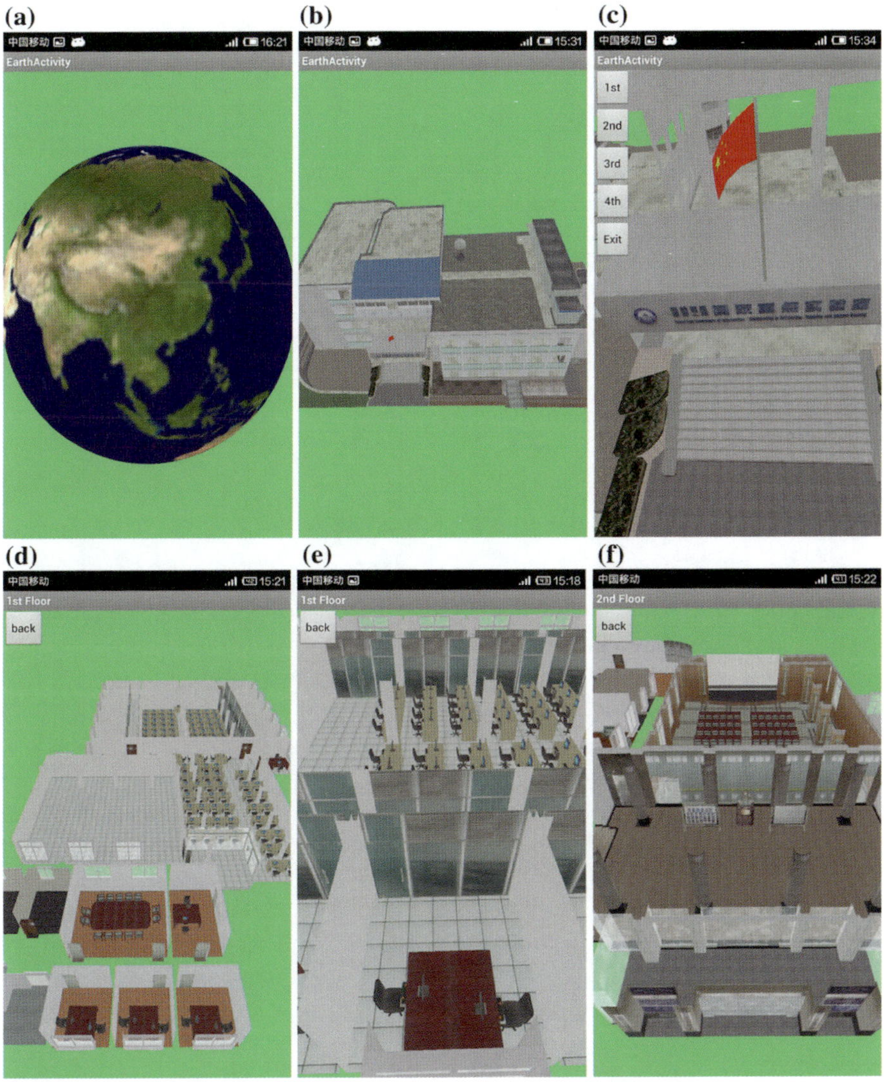

Fig. 1 Different Levels of the 3D Scenes. **a** shows the global scene, **b** is the AOI scene, **c** is the interface of floor choices, **d** shows the single indoor floor scene, **e** and **f** are the details

The AOI Scene is initialized with the shell, only when the user decides to enter the building (conducted by camera collision detection), an interface is prompted to provide choice about which floor model to load, thus maintains the appropriate quality of visualization as well as a reasonable frame rate. To simulate the viewer's perspective, the second and third scenes are set with a fixed camera angle with the horizontal plane.

Mobile GPUs are usually designed to render high-quality images with emphasis of lower power consumption and high speed and thus serves as an ideal remedy for the limited battery and CPU on mobile phones (Akenine-Moller and Strom 2008). Our scene rendering is graphics processing units (GPU) accelerated based on GLSL, a programmable shading language that provides vertex shader and fragment shader in the rendering pipeline. When the application starts, the CPU sends triangles to the GPU to be rendered. The vertex processing first transforms the vertices of a triangle into desired positions. Then three vertices are assembled to a triangle and each pixel inside that triangle are identified and send to the fragment shader. This includes computing the color of the pixel using a variety of techniques such as texture and lighting. At the end of the GPU processing, different types of frame buffer operations take place. This includes resolving visibility and blending. Finally, the result is written to memory.

3 3D Indoor Positioning Based on Smart Phones

With the development of Internet and smart phones, 3D indoor positioning is paid more and more attention. GPS can provide effective outdoor location service (Bajaj et al. 2002), but the GPS signal is often blocked inside the buildings, where people spend most of their time. WiFi, RFID, ultrasound, radio and etc. has been proposed as alternatives to provide indoor positioning services. However, these methods rely on huge pre-investment of devices and therefore are not applicable in public or non-profit buildings.

This paper proposed a new method to locate the user's position without huge pre-investment and tedious procedure. It mainly includes two parts: Accurately intialize the start position using close-range photogrammetry and autonomously track the user's location based on device' embedded sensors.

3.1 Calculate the Initial Position Accurately

One of the bottlenecks for indoor positioning and navigation lies in precise determination of initial position. Winter and Kealy (2012) combined the movement pattern of the user and the geometry structure of the building to determining initial position. However it can not work effectively when the movements are irregular. This paper proposed an novel solution which is insusceptible to the user's movements and can feedback the current position immediately by taking a photo of the pre-deployed cross board using the mobile phone's camera.

This method is based on the principle of Single Photo Resection in Photogrammetry whose main idea is to obtain the image orientation and position using control points, whose coordinate are known both in the image and object reference systems (Tommaselli and Reiss 2005). And the cross board is employed to provide control points informations.

Laser range scanner, combined with the panoramic camera, tend to play an important role in building realistic, visually convincing 3D model of indoor spaces. The densely sampled points collected by laser provide rich location information and can be used to determine the location of control points. We can obtain the location of control points by building the model of indoor spaces with cross boards attached or not. In the former strategy, the control points can be extracted from cloud points based on feature extraction methods (Daniels et al. 2007; Biber et al. 2004). While in the other one, the cross boards will be added to the building later and the position of the control points referred to the model coordinates could be measured using other technology, for example the total station. In our experiment, the former strategy is preferred for its relatively less work.

The steps are as follows:

1. Cross board deployment
Considering the easy-accessibility of the cross board, they should be placed in obvious and important regions, such as the extrance of the building, the gate to a new floor, the gate to get up or down the stairs, and the corner to a different direction. Furthermore, this method is the basis for subsequent correction of Inertial Navigation path, the density of cross board is dependent on the accuracy of Inertial discussed in Sect. 4.2.
2. Control points extraction and image-coordinates calculating

   - We employ Small Univalue Segment Assimilating Nucleus (SUSAN) edge detection method (Smith and Brady 1997) to extract and localize the four corners of the cross board in image processing. The method can acquire features (edge, corner) of objects with precise localization and it is not sensitive for local noise. As show in Fig. 2a, 12 corners of the cross board, marked in clusters of blue points, were first recognized. Then, applying k-means cluster algorithm to identify the center of each cluster (marked in red), which serves as the final location of 12 corners.
   - Predefining a square template with a size 10 × 10 pixels. Applying it to 12 corners and calculating the average intensity of each template. Traversing each pixel of the template, make a statistic about the number of pixels that are greater and lower than the average intensity, and marked by a and b, respectively. If b is greater than a, the corner was considered to be located in the intersection of two lines and thus should be disposed. If the other way, the corner should be reserved for further calculating.
   - The left eight corners are treated as four pairs, corresponding to four end-points. For each pair, a median point was calculated and the coordinate of this point is the image coordinate of control point, as show in Fig. 2b.

The calculation of photo coordinate involves coordinate transformation from photo coordinate to the image coordinate. The origin, scale and axis of the photo

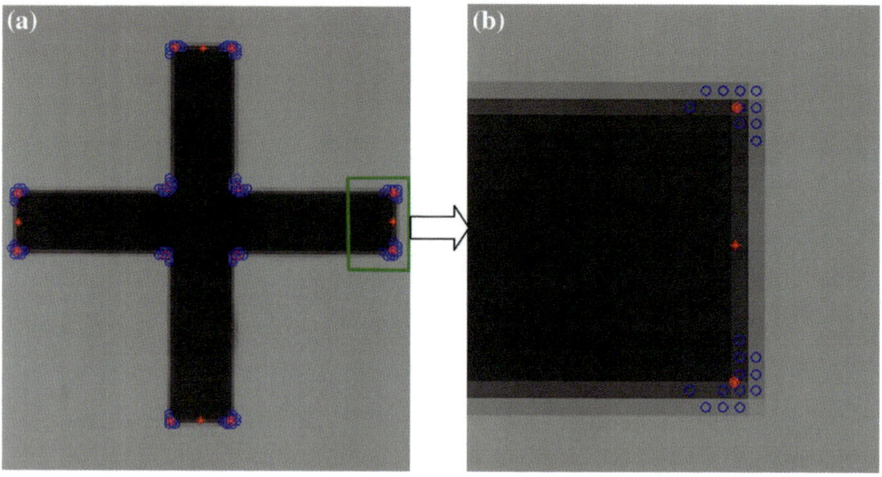

Fig. 2 The extraction of control points

and image coordinate are probably not the same. The transform formula is shown below.

$$x = k(j - col/2) \tag{1}$$

$$y = k(-i + row/2) \tag{2}$$

where (x, y) is the image coordinate of control points, (i, j) is the row and column, row, col are the height and width of the photo respectively, k is the size of every pixel.

3. Initial position calculating

The initial position is calculated based on Single Photo Resection, which is founded on collinearity equation, this is, elements of exterior orientation (X_s, Y_s, Z_s, φ, ω, κ) are calculated based on the control points' photo and object coordinate. Two equations can be listed according to one pair of photo and object coordinate. So six equations can be listed to calculate six exterior orientation elements if three control points are given. This paper uses four control points on the cross board to improve precision by least square adjustment.

The collinearity equation is shown below.

$$\begin{aligned} x &= -f\frac{a_1(X - X_s) + b_1(Y - Y_s) + c_1(Z - Z_s)}{a_3(X - X_s) + b_3(Y - Y_s) + c_3(Z - Z_s)} \\ y &= -f\frac{a_2(X - X_s) + b_2(Y - Y_s) + c_2(Z - Z_s)}{a_3(X - X_s) + b_3(Y - Y_s) + c_3(Z - Z_s)} \end{aligned} \tag{3}$$

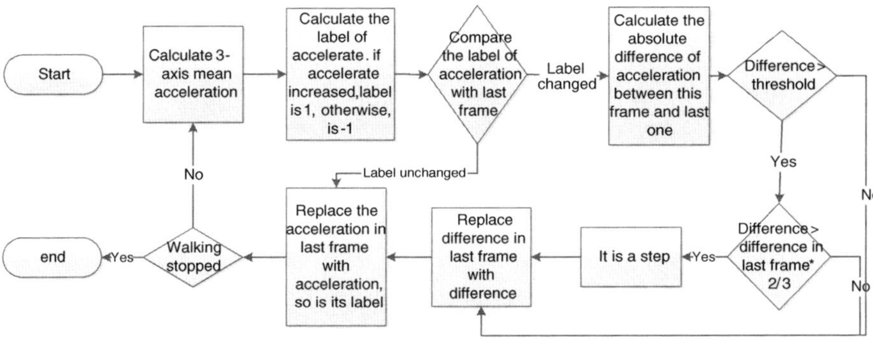

Fig. 3 Pedometer algorithm flowchart

where (x, y) is the Image coordinate and X, Y, Z are the object coordinates of control points, X_s, Y_s, Z_s are the object coordinates we calculate respectively, a_i, b_i, c_i $(i = 1, 2, 3)$ are the direction cosines of φ, ω and κ.

3.2 Calculate Continuous Location Based on Inertial Devices

Inertial navigation system (INS) is an autonomous navigation technique which employs accelerometers and gyroscopes to track the position and orientation of a moving object relative to a known starting point, orientation and velocity. Many scholars have applied it to LBS (Kuo et al. 2013; Winter and Kealy 2012; Woodman 2007). However, the drift caused by senor measurement deviations may accumulate over time and need an efficient calibrating algorithm. Given this, we designed a way to obtain the continuous positioning of users, which is implemented by determining the step counts and the heading of each step.

3.2.1 Pedometer Implementation

One step is determined by extracting the extreme point which has the maximum or minimum acceleration and judging if the difference between the last two extreme points is above the pre-defined threshold. The step length and threshold are user-defined based on different conditions. And the procedure is illustrated in Fig. 3.

3.2.2 Azimuth Calculation

By acquiring the raw sensor data from the accelerometer and magnetic device, rotation matrix R from the global coordinate system to the device's could be calculated. Then based on the rotation matrix R, we calculate the device's orientation,

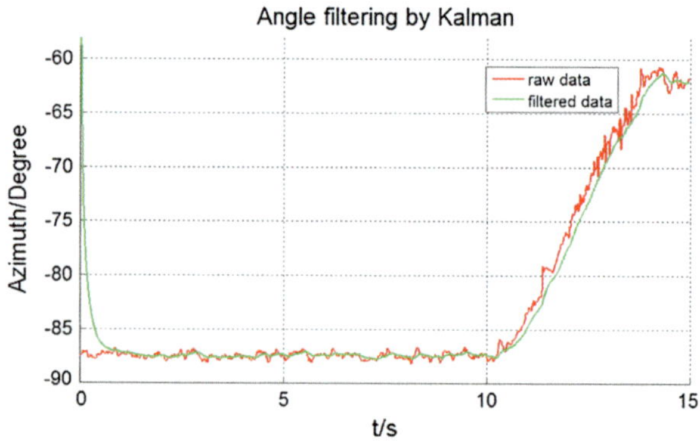

Fig. 4 Filtered data of the azimuth by Kalman. The *red line* shows the raw data gained by phone sensors, which is sensitive to noise, while the *green line* is the data filtered by Kalman

which includes azimuth (rotation around the Z axis), pitch (rotation around the X axis) and roll (rotation around the Y axis).

In order to eliminate the noise of the measurement, we use Kalman filter (Kalman 1960) to improve the accuracy. The Kalman filter is a set of mathematical equations that provides an efficient computational (recursive) solution of the least-squares method. The filter is very powerful in several aspects: it supports estimations of past, present, and even future states, and it can do so even when the precise nature of the modeled system is unknown (Welch and Bishop 2000).

As illustrated in Fig. 4, when following the same direction, the azimuth suffers a fluctuation of 0–2° and grows higher as the device moves or the accuracy of the sensor decreases. However, When processed with Kalman filter, the influence of white noise could be constrained to a lower degree and the track angle can have a higher accuracy, as the green line shows.

4 Experimental Result and Analysis

4.1 Detection of Control Points and Single Photo Resection

In our experiment, a standard 3-axis coordinate system with an origin located 1.67 m below the center of the cross board is used to express the data values. The X axis is horizontal and points to the east, the Y axis is horizontal and points to the south, and the Z axis is vertical to the horizontal plane and points upside. Take photos of the cross board in location P1, P2 and P3, as respectively shown in Fig. 5a–c. Then, extract the control points and calculate their coordinate values based on the methods discussed above. The real positions, calculated positions and

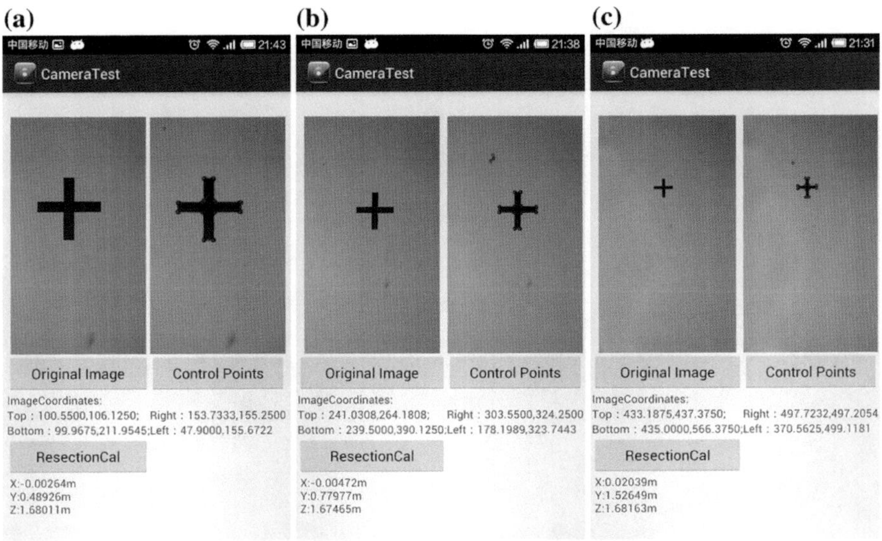

Fig. 5 Location by single photo resection. **a**, **b**, **c** respectively shows the photos taken in position P1, P2, P3, the extraction of feature points with their image coordinate, and the position calculated by single photo resection

residuals are listed in Table 1. The result shows that when the distance between camera and cross board is within 1 m, best performance can be achieved. And the result risks higher inaccuracy as distance increases.

4.2 Integration Navigation Via INS and Single Photo Resection

The real location experiment was carried out on the third floor of our lab, which had a map specified by the Fig. 6. We set off from point A, and walk along the wall with mobile phone in hand to point D. The real trajectory was labeled on the map with black line, while the route calculated by INS was labeled with blue dot line in a of Fig. 6. In the comparison experiment, we corrected the position using close-range photogrammetry at point E, as shown in b and the estimated trajectory is displayed by the red line.

From the experiment result, we can know that when moving forward using inertial navigation, the real path may deviate from the theoretical results, mainly because the direction of mobile phone device does not exactly correspond with the moving direction. The accumulated error, especially long time angle deviation, may lead to a overall mismatch of the trajectory, as the blue dot line shows. Therefore, a correction strategy is of great necessary. As shown in b of Fig. 6, we corrected the point E to F on the trajectory using the close-range photogrammetry, which

Table 1 Accuracy results by close-range photogrammetry

P1			P2			P3		
X(m)	Y(m)	Z(m)	X(m)	Y(m)	Z(m)	X(m)	Y(m)	Z(m)
0	0.4535	1.6853	0	0.7353	1.6257	0	1.4058	1.5102
0.00264	0.48926	1.68011	−0.00228	0.82875	1.67992	0.02039	1.52649	1.68163
0.00264	0.03576	0.00519	0.00472	0.04447	0.05195	0.02039	0.12069	0.17143
0.0362/0.4537 = 7.9788 %			0.06854/0.7366 = 9.3049 %			0.2106/1.4148 = 14.8855 %		

The first row is the real position of P1, P2 and P3. The second row shows the position calculated by the method proposed in this paper. The third row is the residuals of X, Y, Z. And the relative errors of distance are shown in the last row

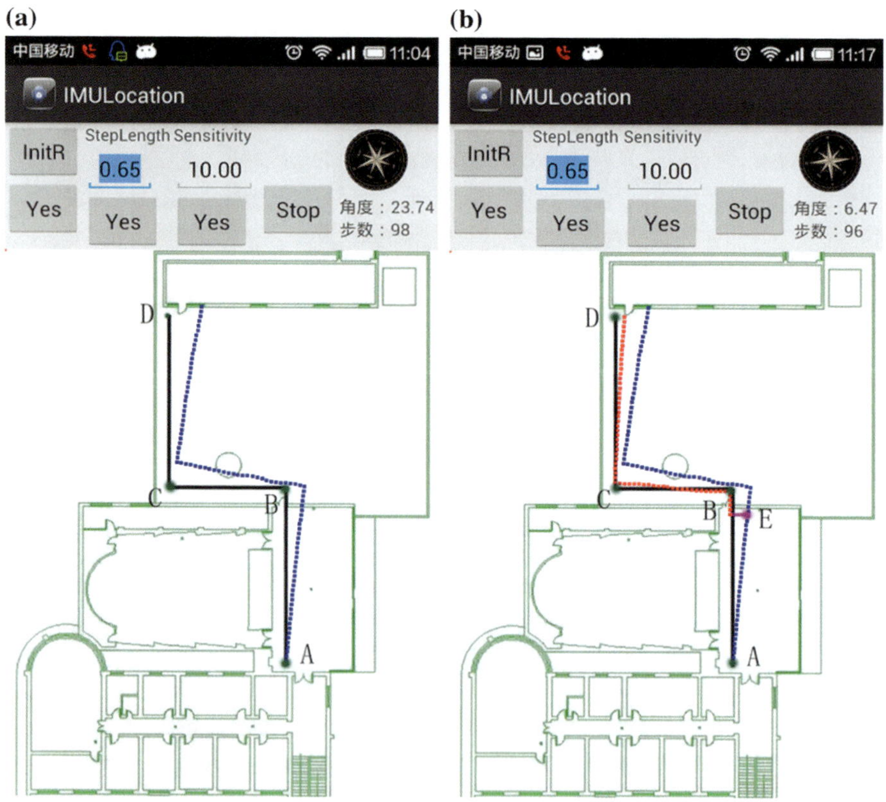

Fig. 6 INS trajectory and its correction. The *black line* represents the truth trajectory of the experiments and the *blue dot line* shows the estimation route without correction during the process. The *red line* in **b** displays the advanced result using the correction at point E

eliminated the accumulated error of the azimuth and aligned the estimated trajectory with the truth better.

Making correction could improve the accuracy of positioning, however it costs a lot. The key to compromise between the two factors is to find the optimal density of the cross boards distribution. The density of cross board remained unsolved in

Sect. 3.1 can be explained here by taking into account the accuracy of inertial navigation. If the system has a required position error that is within 3 m, the accuracy of inertial navigation is 95 %, which means every 60 m risks a deviation of 3 m. So the most appropriate density of cross board is one per 60 m.

5 Conclusion

In this paper, a 3D indoor positioning system is established. It can provide 3D indoor visualization friendly and efficiently based on GPU rendering. The initial position can be calculated accurately according to close-range photogrammetry, and the continuous position can be obtained by sensors embedded in phones. Compared with other ways like WiFi, RFID, ultrasound, radio and etc., this method cost little and can be implemented easily, which makes it more convenient for users to operate. In the future, we will polish the method to extract the control points more automatically and smartly.

Acknowledgement This work was supported by a grant from the National High Technology Research and Development Program of China (863 program) (No. 2012BAH35B03 and No. 2013AA12A203) and a grant from Shenzhen Science and Technology Development Program.

References

Akenine-Moller T, Strom J (2008) Graphics processing units for handhelds. Proc IEEE 96 (5):779–789

Bahl P, Padmanabhan VN (2000) RADAR: an in-building RF-based user location and tracking system. In: INFOCOM 2000. Nineteenth annual joint conference of the IEEE computer and communications societies. Proceedings IEEE vol 2, pp 775–784

Bajaj R, Ranaweera SL, Agrawal DP (2002) GPS: location-tracking technology. Computer 35 (4):92–94

Biber P, Andreasson H, Duckett T, Schilling A (2004) 3D modeling of indoor environments by a mobile robot with a laser scanner and panoramic camera. In: International conference on intelligent robots and systems, 2004. (IROS 2004). Proceedings 2004 IEEE/RSJ (vol 4, pp 3430–3435)

Daniels J, Ha LK, Ochotta T, Silva CT (2007) Robust smooth feature extraction from point clouds. In: IEEE international conference on shape modeling and applications, 2007. SMI'07. (pp 123–136)

González E, Prados L, Rubio AJ, Segura JC, de la Torre Á, Moya JM, Martín JL (2009) ATLINTIDA: a robust indoor ultrasound location system: design and evaluation. In: 3rd symposium of ubiquitous computing and ambient intelligence 2008 (pp 180–190). Springer, Berlin

Kalman RE (1960) A new approach to linear filtering and prediction problems. J Fluids Eng 82 (1):35–45

Kuo YC, Hsiao MY, Wen CY (2013) An integrated mobile sensor platform for collaborative indoor self-positioning applications. In: TENCON spring conference, 2013 IEEE (pp 495–499)

Prasithsangaree P, Krishnamurthy P, Chrysanthis PK (2002). On indoor position location with wireless LANs. In: The 13th IEEE international symposium on personal, indoor and mobile radio communications, 2002. (vol 2, pp 720–724)

Smith SM, Brady JM (1997) SUSAN—a new approach to low level image processing. Int J Comput Vision 23(1):45–78
Welch G, Bishop G (2000) An introduction to the Kalman filter, from http://www.cs.unc.edu, UNC-Chapel Hill, TR95-041, November
Tommaselli AMG, Reiss MLL (2005) A photogrammetric method for single image orientation and measurement. Photogram Eng Remote Sens 71(6):727–732
Winter S, Kealy A (2012) An alternative view of positioning observations from low cost sensors. Comput Environ Urban Syst 36(2):109–117
Woodman OJ (2007) An introduction to inertial navigation. University of Cambridge, Comput Lab Tech Rep UCAMCL-TR-696 14, 15

Range Domain IMM Filtering with Additional Signal Attenuation Error Mitigation of Individual Channels for WLAN RSSI-Based Position-Tracking

Seong Yun Cho

Abstract In this paper, an adaptive filter is presented for position-tracking of a mobile node using WLAN Received Signal Strength Indicator (RSSI). To take the dynamics of the mobile node into consideration, the presented filter is expressed based on an Interacting Multiple Model (IMM) filter. In indoor environment, Additional Signal Attenuation Error (ASAE) occurs due to several obstacles such as wall, furniture, etc. It causes large positioning error. The presented filter includes an ASAE mitigation function of individual channels. In the simulation test, it shows that the presented filter can provide an accurate position-tracking solution for a mobile node using WLAN RSSI in indoor environment.

Keywords WLAN RSSI · Position-tracking · Additional signal attenuation error mitigation

1 Introduction

Position-tracking of a pedestrian using wireless communication signals has been the main research topic with the advent of the location-based services (LBS) market (Kolodziej and Hjelm 2006). In open area, position-tracking solutions can be easily obtained through satellite navigation systems such as global positioning system (GPS). However, there is still a need for using ground-based infra such as WLAN, ultra-wideband or chirp spread spectrum to fill up the radio shadow area of satellite navigation systems.

Recently, WLAN-based indoor position-tracking (WIP) technology has been widely investigated owing to its usefulness that many WLAN access points (AP) have been already installed in most buildings for communications and the WLAN AP signals can be used free for position-tracking. That is the merits of WIP.

S.Y. Cho (✉)
Department of Applied Robotics, Kyungil University, Gyeongsan-si 712-701, South Korea
e-mail: sycho@kiu.kr

G. Gartner and H. Huang (eds.), *Progress in Location-Based Services 2014*,
Lecture Notes in Geoinformation and Cartography,
DOI 10.1007/978-3-319-11879-6_7

However, WLAN is a time asynchronous system. That is, it is difficult to obtain accurate time-of-arrival (TOA) measurements based on the general WLAN AP and modem. So, received signal strength indicator (RSSI) has been used generally in WIP. There are essentially two categories of RSSI-based position-tracking. One uses a fingerprinting database (Cho and Yun 2009; Li et al. 2005). The basis of fingerprinting method is first to establish a database that contains the measured RSSI at some reference points in the service area. Then the location of a mobile node can be identified by comparing its RSSI measurements with the database. This class of technology has received more attention owing to its accuracy recently. However, the disadvantages of this approach are the database generation and maintenance requirements. Practically, it is difficult to establish the fingerprinting databases for indoor spaces of all of buildings.

The other category of WIP is RSSI-based trilateration (Cho 2010; Whitehouse et al. 2007). To achieve this, a RSSI measurement has to be converted to a range measurement using a proper signal propagation model (Cho 2010). Also location information of APs has to be known for trilateration-based localization. Fortunately, it is easy to estimate the location information of APs (Cho 2010). However, indoor radio signal propagation is very complicated, because of signal attenuation due to distance, penetration losses through walls and the effect of multipath propagation. The signal attenuation due to distance can be modeled, whereas the other terms cannot be designed as a model. So, WIP may yield position-tracking solutions with large errors caused by the un-modelled terms. In this paper, these terms are called additional signal attenuation error (ASAE).

Range measurement-based position estimation methods can be divided into two parts. One is a filter-free method and the other is a filter-based method. The filter-free method includes iterative methods and closed-form solutions (Cho and Choi 2009; Cho and Kim 2012). Many filter-free methods are based on the assumption of pure line of sight (LOS) propagation with additive Gaussian white noise (AWGN). However, filter-free methods with AWGN are a mismatch to the physical situation and there would likely be a performance disadvantage in using them in real environment. On the other hand, the filter-based method can use statistical models to mitigate the effect of the ASAE (Ai-Jazzar et al. 2002; Huerta and Vidal 2009; Banani et al. 2013). However, it is difficult to design the exact model of the ASAE. That is, the filter-based method can provide comparatively accurate solution compared with the filter-free method, but estimation performance degradation of the filter cannot be avoided due to the ASAE.

In this paper, an ASAE estimation technique is presented based on the residual of the filter. The basis of the filter is expressed based on an interacting multiple model (IMM) filter (Bar-Shalom et al. 2005) containing a constant velocity (CV) model and a constant acceleration (CA) model (Li and Jilkov 2003) in the light of the mobile node dynamics. In each sub-filtering with a different model, the residual is used to mitigate the ASAE before processing the measurement-update. The motivation of this work is that the residual in epoch-by-epoch position-tracking contains the current ASAE. Using the statistical property of the ASAE, the ASAE estimates of individual channels can be extracted and the measurements can be compensated. Then the measurement-update is processed using the compensated measurements.

The performance of the presented filter is evaluated by simulation. In the simulation approach, the results of the presented filter is compared with the iterative least squares (ILS) method representing the filter-free method and the extended Kalman filter representing the filter-based method. From this simulation result, it shows that the presented filter can greatly enhance the performance of WIP.

The rest of the paper is organized as follows. Section 2 describes WIP and range domain filter for WIP. The range domain IMM filter with ASAE mitigation of individual channels is presented in Sect. 3 with the simulation results in Sect. 4. Finally, Sect. 5 gives conclusions.

2 WLAN-Based Indoor Position-Tracking

Position-tracking problem is how to estimate the position of a mobile node by using available measurements. Two main cores to solve the problem are the measurement and estimation method. In this section, filter-based position-tracking method using WLAN RSSI measurements is explained.

2.1 Radio Propagation Model and RSSI-Based Range Measurement

The power of the propagating signal is attenuated through passing air and is modeled as (Cho 2010)

$$\begin{aligned} \tilde{P}(r) &= P(1) - 10\alpha \log_{10}(r) - \delta P(r) + w_r, \\ &\equiv P(r) - \delta P(r) + w_r \end{aligned} \tag{1}$$

where$\tilde{P}(r)$ denotes the RSSI measurement [dBm] obtained in a mobile node located from an AP with distance of $r.\alpha$ denotes the attenuation factor in the free space, δP denotes signal strength level ASAE caused by wall penetration, multipath signals and non-line-of-sight (NLOS) error and w_r denotes signal strength level AWGN.

$P(1)$ and α can be estimated using the measurements in LOS space as

$$\begin{bmatrix} \hat{P}(1) \\ \hat{\alpha} \end{bmatrix} = (M^T M)^{-1} M^T Y, \tag{2}$$

where

$$M = \begin{bmatrix} 1 & -10\log_{10}(r_1) \\ 1 & -10\log_{10}(r_2) \\ \vdots & \vdots \\ 1 & -10\log_{10}(r_m) \end{bmatrix}, \tag{3}$$

$$Y = \begin{bmatrix} \widetilde{P}(r_1) & \widetilde{P}(r_2) & \cdots & \widetilde{P}(r_m) \end{bmatrix}^T, \tag{4}$$

where m is the number of the acquired data and r_j is the signal acqusition distance between an AP and the mobile node.

Using (1) and (2), RSSI-based range measurement equation at time k can be denoted as

$$\tilde{r}_k = 10^{\tilde{\beta}_k}, \tag{5}$$

where

$$\tilde{\beta}_k = \frac{\hat{P}(1) - \tilde{P}_k - \delta P_k + w_k}{10\hat{\alpha}}. \tag{6}$$

If the ASAE cannot be estimated (6) can be calculated practically as

$$\tilde{\beta}_k = \frac{\hat{P}(1) - \tilde{P}_k}{10\hat{\alpha}}. \tag{7}$$

2.2 Range Domain Filter for WIP

To design the WIP filter, it is difficult to set a single system model for representing the walking dynamics of a pedestrian containing a mobile node. However, the following CV model has been used (Li and Jilkov 2003)

$$X_{k+1}^{CV} = F^{CV} X_k^{CV} + w_k$$
$$\Leftrightarrow \begin{bmatrix} x_{k+1}^m \\ \dot{x}_{k+1}^m \\ y_{k+1}^m \\ \dot{y}_{k+1}^m \end{bmatrix}^{CV} = \begin{bmatrix} 1 & T & 0 & 0 \\ 0 & 1 & 0 & 0 \\ 0 & 0 & 1 & T \\ 0 & 0 & 0 & 1 \end{bmatrix}^{CV} \begin{bmatrix} x_k^m \\ \dot{x}_k^m \\ y_k^m \\ \dot{y}_k^m \end{bmatrix}^{CV} + w_k, w_k \sim N(0, Q^{CV}), \tag{8}$$

where X denotes the state vector, $[x_k^m \ y_k^m]^T$ denotes the position of a mobile node at time k, $\dot{u}$ denotes the velocity state variable of $u \in \{x, y\}$, F denotes the system matrix and T denotes the localization sampling duration that can be set equal to the measurement taking period. Q^{CV} denotes the process noise covariance matrix in the CV model-based filter.

RSSI-based range measurement Eq. (5) can be rewritten as

$$\begin{aligned} \tilde{r}_k^j &= 10^{\tilde{\beta}_k} \\ &= \sqrt{(x^j - x_k^m)^2 + (y^j - y_k^m)^2} + \delta_k^j + w_k^j, \\ &= r_k^j + \delta_k^j + w_k^j \end{aligned} \tag{9}$$

where $[x^j \ \ y^j]^T$ denotes the position of an access point j, δ_k^j and w_k^j denote the range level ASAE and AWGN converted into range measurement, respectively.

Using (9), measurement equation for filtering can be written as

$$\begin{aligned} z_k &= H_k^{CV} \delta X_k^{CV} + v_k \\ \Leftrightarrow z_k &= \begin{bmatrix} -\frac{x^1 - \hat{x}_k^m}{\hat{r}_k^1} & 0 & -\frac{y^1 - \hat{y}_k^m}{\hat{r}_k^1} & 0 \\ \vdots & \vdots & \vdots & \vdots \\ -\frac{x^n - \hat{x}_k^m}{\hat{r}_k^n} & 0 & -\frac{y^n - \hat{y}_k^m}{\hat{r}_k^n} & 0 \end{bmatrix}^{CV} \delta X_k^{CV} + v_k, v_k \sim N(0, R), \end{aligned} \tag{10}$$

where δX denotes the error state vector, n denotes the number of access points, R denotes the measurement noise covariance matrix and H denotes the measurement matrix and

$$\hat{r}_k^j = \sqrt{(x^j - \hat{x}_k^m)^2 + (y^j - \hat{y}_k^m)^2}. \tag{11}$$

The measurement noise is not practically white Gaussian noise due to the ASAE. Therefore, the position-tracking filter may yield estimation errors.

Based on the system and measurement matrices denoted in (8) and (10), the following measurement-update and time-propagation of the state variables and state error covariance matrix are iterated whenever a set of range measurements is obtained.

$$\hat{X}_k = \hat{X}_k^- + P_k^- H_k^T (H_k P_k^- H_k^T + R)^{-1} \zeta_k, \tag{12}$$

$$P_k = (I_{4\times4} - P_k^- H_k^T (H_k P_k^- H_k^T + R)^{-1} H_k) P_k^-, \tag{13}$$

$$P_{k+1}^- = F P_k F^T + Q, \tag{14}$$

$$\hat{X}_{k+1}^- = F \hat{X}_k, \tag{15}$$

where superscript—denotes the time-propagation, P_k denotes state error covariance matrix at the k-th epoch and ζ_k denotes a residual vector calculated as (Brown and Hwang 1985)

$$\zeta_k = R_k - \hat{R}_k\big|_{X=\hat{X}_k^-} - H_k^{CV}\delta X_k^{CV-}. \tag{16}$$

The error state variables are not time-propagated, that is δX_k^{CV-} is equal to a zero vector. Therefore, (16) can be rewritten as

$$\begin{aligned}\zeta_k &= R_k - \hat{R}_k\big|_{X=\hat{X}_k^-}\\ &= \begin{bmatrix}\tilde{r}_k^1\\ \vdots\\ \tilde{r}_k^n\end{bmatrix} - \begin{bmatrix}\hat{r}_k^1\\ \vdots\\ \hat{r}_k^n\end{bmatrix}.\end{aligned} \tag{17}$$

The filter-based position-tracking method can provide relatively accure solutions compared with the filter-free position-tracking methods because the filter estimates position-tracking solutions relying on the dynamic model as well as the measurements.

3 IMM Filtering with ASAE Mitigation of Individual Channels

One of the problems in the position-tracking filter is the ASAE. To mitigate the ASAE, the property of the ASAE is analyzed. Radio signal is attenuated by wall penetration, multipath signals and NLOS error. Signal attenuation causes the increase in ranging. That is, the range level ASAE δ_k^j in (9) is always a positive real value.

Inserting (9) and (11) into (17) yields

$$\begin{aligned}\zeta_k &= \begin{bmatrix} r_k^1 + \delta_k^1 + w_k^1\\ \vdots\\ r_k^n + \delta_k^n + w_k^n\end{bmatrix} - \begin{bmatrix} r_k^1 + \delta r_k^1\\ \vdots\\ r_k^n + \delta r_k^n\end{bmatrix}\\ &= \begin{bmatrix}\delta_k^1 + w_k^1 - \delta r_k^1\\ \vdots\\ \delta_k^n + w_k^n - \delta r_k^n\end{bmatrix}\end{aligned}, \tag{18}$$

where δr_k^j denotes range estimation error caused by the position-tracking error.

From (18), the ASAE of individual channels can be written as

$$\delta_k^j = \zeta_k^j - w_k^j + \delta r_k^j. \tag{19}$$

(19) can be rewritten as

$$\begin{aligned} E[\delta_k^j] &= \hat{\delta}_k^j \\ &= E[\zeta_k^j] - E[w_k^j] + E[\delta r_k^j] \end{aligned}, \tag{20}$$

In (20), $E[w_k^j] = 0$ and $E[\delta r_k^j]$ can converge into near zero if position-tracking filter is completely observable. Because the range level ASAE is positive, (20) can be rewritten as

$$\hat{\delta}_k^j = E[\zeta_k^j] = |\zeta_k^j|. \tag{21}$$

Therefore, the ASAE of individual channels can be estimated using the absolute value of the individual components of the filter residual. Before processing the measurement-update in the position-tracking filter, the residual is purified by the estimated ASAE as

$$\bar{\zeta}_k = \begin{bmatrix} \tilde{r}_k^1 - \hat{\delta}_k^1 - \hat{r}_k^1 \\ \vdots \\ \tilde{r}_k^n - \hat{\delta}_k^n - \hat{r}_k^n \end{bmatrix}. \tag{22}$$

Another problem in the position-tracking filter is a mismatch between the system model that represents the dynamics of a mobile node and the true dynamics of the mobile node. The CV model can be used as a main model. However, the speed or moving direction of a mobile node is changed quickly, the estimation error may occur due to the model mismatch. To solve this problem, IMM filter with two models, CV and CA models, is used in this paper. The CA model-based system and measurement equations can be written as (Li and Jilkov 2003)

$$\begin{aligned} X_{k+1}^{CA} &= F^{CA} X_k^{CA} + w_k \\ \Leftrightarrow \begin{bmatrix} x_{k+1}^m \\ \dot{x}_{k+1}^m \\ \ddot{x}_{k+1}^m \\ y_{k+1}^m \\ \dot{y}_{k+1}^m \\ \ddot{y}_{k+1}^m \end{bmatrix}^{CA} &= \begin{bmatrix} 1 & T & \frac{T^2}{2} & 0 & 0 & 0 \\ 0 & 1 & T & 0 & 0 & 0 \\ 0 & 0 & 1 & 0 & 0 & 0 \\ 0 & 0 & 0 & 1 & T & \frac{T^2}{2} \\ 0 & 0 & 0 & 0 & 1 & T \\ 0 & 0 & 0 & 0 & 0 & 1 \end{bmatrix}^{CA} \begin{bmatrix} x_k^m \\ \dot{x}_k^m \\ \ddot{x}_k^m \\ y_k^m \\ \dot{y}_k^m \\ \ddot{y}_k^m \end{bmatrix}^{CA} + w_k, w_k \sim N(0, Q^{CA}), \text{ and} \end{aligned} \tag{23}$$

$$
\begin{aligned}
z_k &= H_k^{CA}\delta X_k^{CA} + v_k \\
\Leftrightarrow z_k &= \begin{bmatrix} -\dfrac{x^1-\hat{x}_k^m}{\hat{r}_k^1} & 0 & 0 & -\dfrac{y^1-\hat{y}_k^m}{\hat{r}_k^1} & 0 & 0 \\ \vdots & \vdots & \vdots & \vdots & \vdots & \vdots \\ -\dfrac{x^n-\hat{x}_k^m}{\hat{r}_k^n} & 0 & 0 & -\dfrac{y^n-\hat{y}_k^m}{\hat{r}_k^n} & 0 & 0 \end{bmatrix}^{CA} \delta X_k^{CA} + v_k, v_k \sim N(0,R).
\end{aligned} \tag{24}
$$

A CV model-based EKF and a CA model-based EKF are processed individually. After processing each measurement-update, the two sub-filters are fused based on the residuals. First, a mode probability matrix is updated using the likelihood ratios of the sub-filters. The likelihood radio is calculated based on the residuals (Bar-Shalom et al. 2005).

$$
\mu_k = \begin{bmatrix} \lambda_k^{CV} c_{k-1}^{CV} / (\lambda_k^{CV} c_{k-1}^{CV} + \lambda_k^{CA} c_{k-1}^{CA}) \\ \lambda_k^{CA} c_{k-1}^{CA} / (\lambda_k^{CV} c_{k-1}^{CV} + \lambda_k^{CA} c_{k-1}^{CA}) \end{bmatrix} \equiv \begin{bmatrix} \mu_k^{CV} \\ \mu_k^{CA} \end{bmatrix}, \tag{25}
$$

where c^j denotes the normalization factor of the sub-filter j and can be calculated as

$$
\begin{bmatrix} c_{k-1}^{CV} \\ c_{k-1}^{CA} \end{bmatrix} = Mt^T \mu_{k-1}, \tag{26}
$$

where Mt denotes the Markov transition matrix. Mt and μ_0 are set in initialization step.

In (25), λ_k^j denotes the likelihood ratio of the sub-filter j and can be calculated as

$$
\lambda_k^j = \frac{1}{\sqrt{2\pi \| C_k^j \|}} \exp\left\{ -\frac{1}{2} (\bar{\zeta}_k^j)^T (C_k^j)^{-1} \bar{\zeta}_k^j \right\}, \tag{27}
$$

where C_k^j denotes the residual covariance matrix of the sub-filter j and is calculated as

$$
C_k^j = H_k^j P_k^j (H_k^j)^T + R^j. \tag{28}
$$

It is assumed that the sequences of the purified residuals are white Gaussian with zero-mean.

Using the updated mode probability matrix, the state vector and error covariance matrices of the sub-filters are mixed as

$$
\begin{bmatrix} \bar{X}_k^{CV} \\ \bar{X}_k^{CA\_PV} \end{bmatrix} = Mx_k^T \begin{bmatrix} \hat{X}_k^{CV} \\ \hat{X}_k^{CA\_PV} \end{bmatrix}, \tag{29}
$$

$$\begin{bmatrix} \bar{P}_k^{CV} \\ \bar{P}_k^{CA\_PV} \end{bmatrix} = Mx_k^T \begin{bmatrix} P_k^{CV} + [\hat{X}_k^{CV} - \bar{X}_k^{CV}][\hat{X}_k^{CV} - \bar{X}_k^{CV}]^T \\ P_k^{CA\_PV} + [\hat{X}_k^{CV\_PV} - \bar{X}_k^{CV\_PV}][\hat{X}_k^{CV\_PV} - \bar{X}_k^{CV\_PV}]^T \end{bmatrix}, \tag{30}$$

where Mx denotes a mixing probability that is calculated as

$$Mx = \begin{bmatrix} m^{11}\mu_k^{CV}/c_k^{CV} & m^{12}\mu_k^{CV}/c_k^{CA} \\ m^{21}\mu_k^{CA}/c_k^{CV} & m^{22}\mu_k^{CA}/c_k^{CA} \end{bmatrix}, \tag{31}$$

where m^{ij} denotes the element of the Markov transition matrix.

In (29) and (30), $X_k^{CA\_PV}$ and $P_k^{CA\_PV}$ denote the position and velocity parts in X_k^{CA} and P_k^{CA}, respectively. These are required in mixing process because the system dimension of the CV model is different from that of the CA model.

The mixed state vectors and error covariance matrices are redistributed to each sub-filter. The final estimate at current time is calculated as

$$\hat{X}_k = \bar{X}_k^{CV}\mu_k^{CV} + \bar{X}_k^{CA\_PV}\mu_k^{CA}. \tag{32}$$

4 Simulation Results

To evaluate the performance of the proposed filter-based position-tracking, simulation is performed. It is assumed that a pedestrian possesses a mobile node and walks in a working space as shown in Fig. 1. There are 4 access points. In five circles, a pedestrian stops walking for 10 s. Using the PHY characteristics of WLAN and NLOS error properties, signal strength level ASAE and AWGN are set as

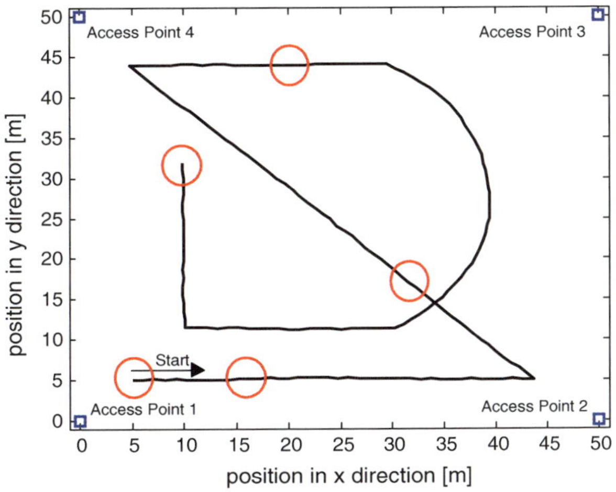

Fig. 1 Working space with 4 access points

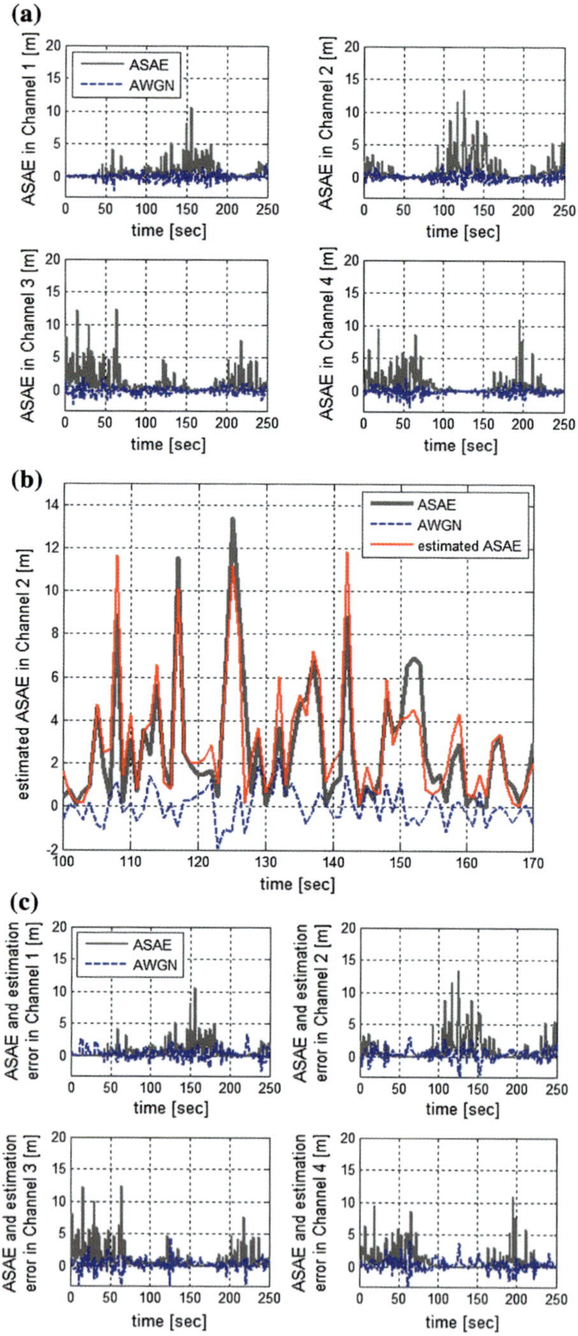

Fig. 2 ASAE and estimates. **a** ASAE and AWGN **b** Estimated ASAE **c** ASAE and estimation error

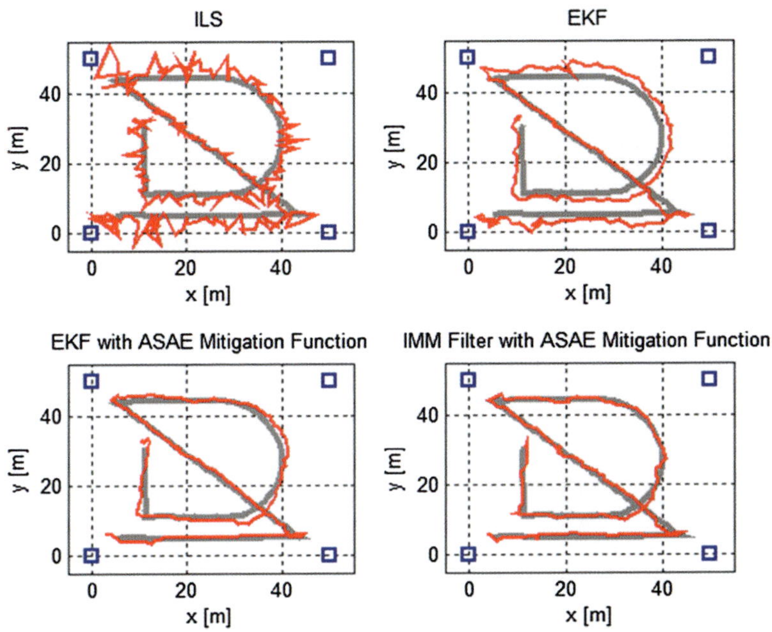

Fig. 3 True trajectory and estimated positions using the ILS, EKF, EKF with ASAE mitigation function and presented filter

$$\delta P(r) = \sum_{i=1}^{2} (0.01 * r * randn)^2, \tag{33}$$

$$w_r = randn/5.0, \tag{34}$$

where *randn* denotes normally distributed random numbers with zero-mean and standard deviation one.

T is set equal to 1.0 s and mean speed of a mobile node is set equal to 1.0 m/s. The initial position of each filter is set by the solution of the ILS method at the first step.

The presented filter is compared with the ILS, EKF with CV model and EKF with ASAE mitigation function. Figures 2 and 3 show the comparison of the estimated ASAEs and positions using the ILS, EKF, EKF with ASAE mitigation function and presented filter.

As shown in Fig. 2a, ASAE is far bigger than AWGN of course and is increased with distance between an AN and mobile node. In this situation, the presented filter estimates the ASAE as denoted in Fig. 2b. It shows that the ASAE estimates of the presented filter epoch-by-epoch position-tracking are close to the true ASAE. To confirm the performance of the ASAE mitigation function, the estimation errors are of individual channels denoted in Fig. 2c.

By comparing Fig. 2a, c, it can be confirmed that the ASAE estimation error is nearly AWGN level. That is, ASAE compensated range measurement-filter can estimate the state variables accurately. The ASAE estimation errors of individual channels are summarized in Table 1. Here STD means standard deviation.

Figure 3 shows the position-tracking results using different methods. As shown in this figure, it can be confirmed that the solutions of the ILS are seriously affected by the ASAE that has a tendency denoted in Fig. 2. Also it can be seen that the EKF provides solutions insensitive to the ASAE. However, the solutions look as if the solutions have bias errors. This phenomenon is also caused by the ASAE. The ASAE can be mitigated by the presented method. The effect is shown in the bottom left plot of Fig. 3. It shows the solutions of the EKF with ASAE mitigation function. The performance has shown a significant improvement compared with the solutions of the EKF. However, in the section after turning the moving direction of a mobile node, the estimation error is slightly increased due to the mismatch between the dynamic model in the filter and the true dynamics of a mobile node. The bottom right plot of Fig. 3 shows the solutions of the presented filter in this

Table 1 ASAE estimation errors

	Channel 1	Channel 2	Channel 3	Channel 4
Mean [m]	0.359	0.403	0.287	0.347
STD [m]	0.813	0.948	0.928	0.958

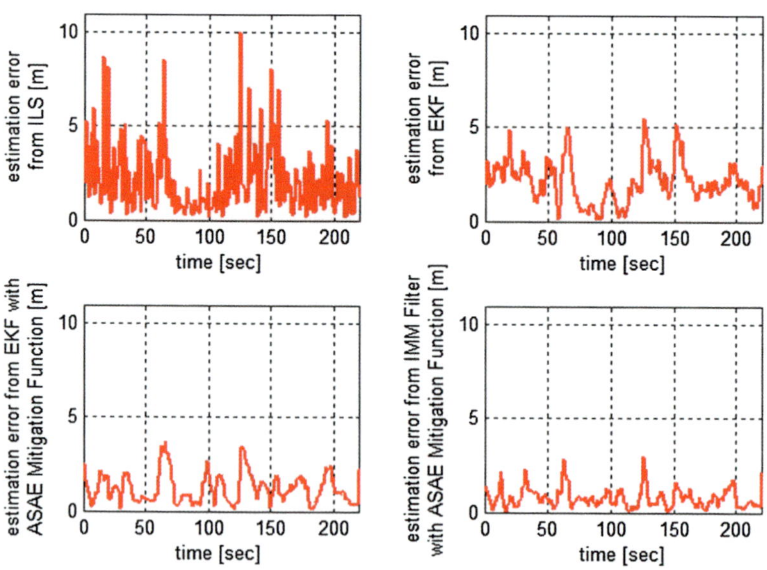

Fig. 4 Position estimation errors

Table 2 Position estimation errors

	ILS	EKF	EKF with AMF	Presented filter
Mean [m]	2.044	2.093	1.275	0.779
STD [m]	1.713	1.054	0.799	0.512

paper. In this figure, it can be confirmed that the proposed filter is nearly free from the ASAE. Also, the mismatch between the dynamic model in the filter and the true dynamics of a mobile node is decreased evidently.

Figure 4 shows the position estimation errors from different methods and Table 2 summarized the comparison of the position-tracking performance among the methods. Consequently, it can be concluded that the proposed filter can practically enhance the performance of the WLAN-based wireless position-tracking in indoor situations.

5 Conclusion

In this paper, a filter-based position-tracking technology using the WLAN RSSI is presented. In indoor environment, there are several obstacles that hinder radio signal from propagating directly between an AP and a mobile node. This situation causes additional signal attenuation in the received signals. That is, RSSI-based position-tracking solutions have large errors in indoor environment. To solve this problem, an ASAE mitigation technology is presented based on the ASAE characteristics and filter residual. Also an IMM filter containing CV and CA models is used in the basis of the position-tracking filter to avert any error growth considering the mismatch between the dynamic model in the filter and the true dynamics of a mobile node. The performance of the proposed filter is analyzed by simulation. The results of the simulation show that WLAN RSSI-based indoor position-tracking technology can be advanced by the filter proposed in this paper.

References

Ai-Jazzar S, Caffery J, You H (2002) A scattering model based approach to NLOS mitigation in TOA location systems. IEEE Veh Technol Conf 2:861–865

Banani S, Najibi M, Vaughan R (2013) Range-based localisation and tracking in non-line-of-sight wireless channels with gaussian scatterer distribution model. IET Commun 7(18):2034–2043

Bar-Shalom Y, Challa S, Blom H (2005) IMM estimator versus optimal estimator for hybrid systems. IEEE Trnas Aerosp Electron Syst 41(3):986–991

Brown R, Hwang P (1985) Introduction to random signals and applied kalman filtering. Wiley, New York

Cho S (2010) Localization of the arbitrary deployed APs for indoor wireless location-based applications. IEEE Trans Consum Electron 56(2):532–539

Cho S, Choi Y (2009) Access point-less wireless location method based on peer-to-peer ranging of impulse radio ultra-wideband. IET-Radar Sonar Navig 4(5):733–743. doi:10.1049/iet-rsn.2009.0157

Cho S, Kim B (2012) Linear closed-form solution for wireless localisation with ultra-wideband/chirp spread spectrum signals based on difference of squared range measurements. IET-Wireless Sens Syst 3(4):255–265. doi:10.1049/iet-wss.2012.0159

Cho S, Yun S (2009) Efficient fingerprint db generation method for indoor wireless location using the environment analysis Tool. ION 2009 International Technical Meeting pp 793–797

Huerta J, Vidal J (2009) Joint particle filter and UKF position tracking in severe non-line-of-sight situations. IEEE J Sel Top Sign Proces 3(5):874–888

Kolodziej K, Hjelm J (2006) Local positioning systems: LBS applications and services. Taylor & Francis Group, Baco Raton, FL

Li X, Jilkov V (2003) Survey of maneuvering target tracking. part I: dynamic models. IEEE Trans Aerosp Electron Syst 39(4):1333–1364

Li B, Wang Y, Lee H, Dempster A, Rizos C (2005) Method for yielding a database of location fingerprints in WLAN. IEE Proc-Commun 152(5):580–586

Whitehouse K, Karlof C, Culler D (2007) A practical evaluation of radio signal strength for ranging-based localization. Mob Comput Commun Rev 11(1):41–52

Application of the Inertial Navigation System 3D-Self-Calibration-Method for the Minimization of the Measurement Uncertainty

Enrico Köppe, Daniel Augustin, Tabea Wilk, Andreas Subaric-Leitis, Achim Liers and Jochen Schiller

Abstract For the accuracy of inertial navigation systems for indoor localization it is important to get high quality sensor data of the multi-sensor system. This can be realized using high quality sensors or the developed 3D-self-calibration-method for low cost sensors. Based on the calibration procedure of the accelerometer (ACC) and the magnetic field sensor (MAG), the additional integration of the gyroscope (GYRO) leads to a reduction of the indoor positioning error. This improves both the approximation for the accelerometer, the magnetic field sensor and the gyroscope so that the standard deviation of a single sensor is minimized. There are errors in the whole system. To determine these error sources it is important to define the measurement uncertainty. In this paper it is presented that the measurement uncertainty can be reduced by the application of the developed 3D-self-calibration method.

Keywords 3D sensor · 3D calibration method · Indoor localization · Measurement uncertainty

1 Introduction

Inertial Navigation Systems gained more and more interest in the last years. Considerable companies as Google Inc., Apple Inc. and other companies and institutes started projects for indoor localization with the Inertial Navigation System. Such projects are: Mircosoft (2014), Google (2014) and DLR (2013). These systems base on expensive sensors. In a project with the FU Berlin the aim was to develop a system with different 3D low cost sensors. This was realized in a dissertation: Köppe (2014), the system is called BodyGuard-System. During the

E. Köppe (✉) · T. Wilk · A. Subaric-Leitis
BAM Federal Institute for Material Research and Testing, Berlin, Germany
e-mail: enrico.koeppe@bam.de

D. Augustin · A. Liers · J. Schiller
Department of Mathematics and Computer Science, FU-Berlin, Berlin, Germany

G. Gartner and H. Huang (eds.), *Progress in Location-Based Services 2014*,
Lecture Notes in Geoinformation and Cartography,
DOI 10.1007/978-3-319-11879-6_8

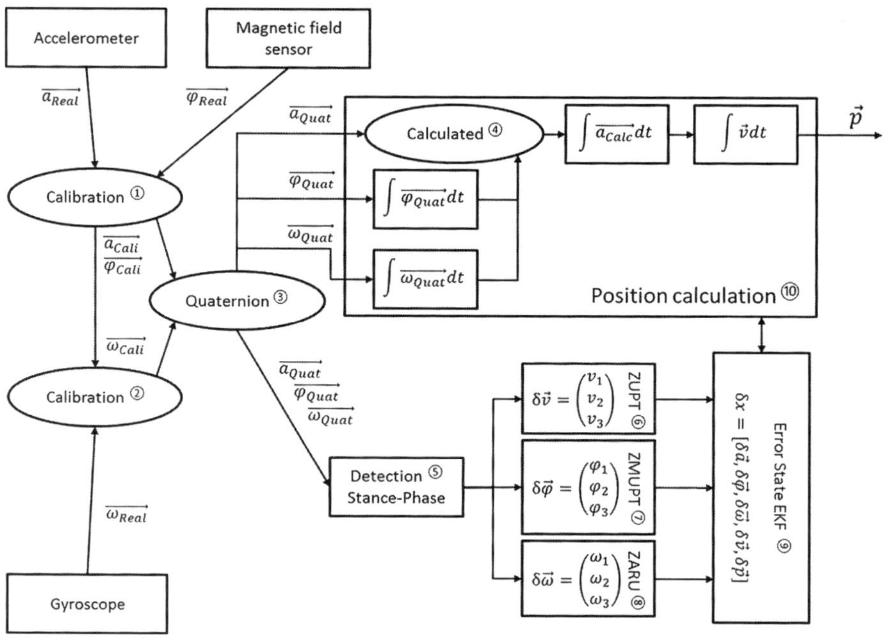

Fig. 1 Analysis of the measured data with the ACC, MAG and GYRO for the position calculation with possible sources of error

development of the system the problem appeared that the low cost sensors are not well calibrated (linearity 2 %, noise rate, scale factor, temperature influence and more). To improve the results of the 3D indoor localization it was necessary to develop a calibration method.

The indoor localization algorithm of the BodyGuard-System is shown in Fig. 1 [derived from Fig. 7.11 of Schmid (2012)]. The analysis includes the calibration procedure and the calculations necessary for the determination Köppe et al. (2014) of the actual position. In Fig. 1 the different analysis steps are numbered from ① up to ⑩. The first and second step are the calibration of the ACC, the MAG and the GYRO. Then the mathematical analysis and the error correction are performed. The error state extended Kalman filter (EKF) is based on the strapdown-algorithm (Titterton 2004). The error equation in Fig. 1 step number ⑨ for the calculation of the position (state vector) δx consists of the five 3D vectors: acceleration $\delta\vec{a}$, angle velocity $\delta\vec{\varphi}$, rotation $\delta\vec{\omega}$, speed $\delta\vec{v}$ and position $\delta\vec{p}$. The advantage using the errors in the error state EKF in comparison to the direct use of measurement result is that there is no need for a linearization. Finally the position $\vec{p}$ is calculated.

The aim of this work is to show that the single uncertainty components can be minimized by the application of the 3D-self-calibration-method. First in the next sections the calibration algorithm and measurement uncertainty will be explained. As a result the improvements in the localization will be shown.

2 3D-Self-calibration-Method

The 3D-self calibration-method was developed to minimize measurement uncertainties resulting of the used low cost sensors. The method is free of external equipment. In the BodyGuard-System (Köppe 2012) three 3D sensors are used: the acceleration sensor, magnetic field sensor and the gyroscope. These three sensors are continuously recalibrated in the method.

In the following the calibration of each single sensor is described and then in the last section the whole method explained. The single sensors are adjusted using the dynamic calibration. Physically the dynamic calibration with the BodyGuard-System is realized by a "slow" turning respectively by the normal movement of a person (Köppe et al. 2012). During the movement the important values for the calculation are determined. These values are the particular zero-points of the accelerometer, the magnetic field sensor and the gyroscope as well as the acceleration values for 1 g, the magnetic field strength, the orientation of the local magnetic field and the angular velocity. Afterwards the determined values and the method of least squares are used to mathematical image the values of the accelerometer and magnetic field sensor as scatter plot on the curved surface area of an ellipsoid. In the process the theory of the asymmetric (real) movement model is used. For each sensor (accelerometer or magnetic field sensor) an own ellipsoid is formed. Each ellipsoid is determined by a vector (r_x, r_y, r_z) and a center point $M(x_{0,}, y_0, z_0)$. The measured values of the gyroscope are shown as line because it is a rotational speed.

This theoretical approach is the basis for the calibration of the single sensors which will be discussed shortly in the next three sections. In the fourth section the whole calibration process is explained.

2.1 Necessary Localization Sensors

For the calibration of the accelerometer and magnetic field sensor the approximated ellipsoid using the dynamic calibration procedure as explained before is shown in Fig. 2. In this figure the scatter plot of all measured values (red dots/lines) are shown. For the accelerometer the projection of the acceleration values is determined in Fig. 2a, on the xy-level of the X and Y axis (black dots/lines), the yz-level of the Y and Z axis (blue dots) and the xz-level of the X and Z axis (green dots). For the magnetic field sensor the projection of the magnetic field values is illustrated in Fig. 2b on the xy-level in z-direction (black line), the yz-level in x-direction (blue line) and the xz-level in y-direction (green line).

The radii or rather the measured values of the single axes are determined with $r_x = 1,048$, $r_y = 998$ and $r_z = 1,064$. These radii are in conformity with the calibration values of 1 g of the gravitation field of the earth for every single axis. The deviation of the radius is about ±40 points compared to the values of the data sheet (1,024 points) as well as about ±10 points compared to each single axis with the

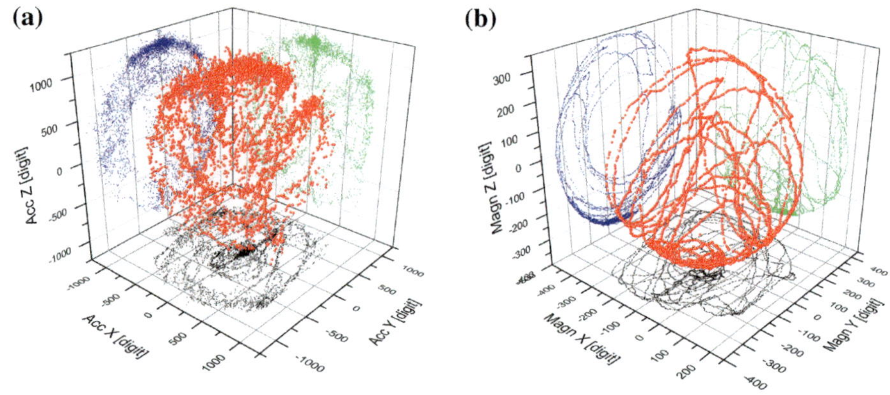

Fig. 2 **a** Ellipsoid of the accelerometer; **b** Ellipsoid of the magnetic field sensor

unique static calibration or manufacturer's calibration. Additionally the zero-point 0 g is determined for each axis. This zero-point with its offset is the center point of the ellipsoid with $r_{xOffset} = 10$, $r_{yOffset} = 0$ and $r_{zOffset} = 4$.

With the radii approximation of the magnetic field sensor the transfer to the curved surface can be nicely demonstrated. The radius determined on the curved surface is only influenced by the external magnetic field of the earth. This influence is described as "hard-iron offset" and "soft-iron offset". It is visible by the displacement of the zero-point. The calculation of the offsets can be done with the software for the extrapolation by the shift of the radii. After the determination of the parameters the values for the mapping of the magnetic field sensor with the radii of the curved surface $r_x = 248$, $r_{xOffset} = -100$, $r_y = 328$ and $r_z = 356$ can be used for the calibration.

For the dynamic calibration free movements are used to determine the values of the gyroscope. The measured values of the gyroscope are approximated in a "curved line" for the angle velocity. This is shown in Fig. 3. Due to the transformation of the sensor values by the relative movement of the three rotation axes the values of the gyroscope are plotted as angle velocity in °/s on the curved line. In the figure we see a red thick line in the xy-level for the yaw axis, a blue thick line in the yz-level for the roll axis and a green thick line in the xz-level for the pitch axis. The thicker lines show the real gyroscope values for the angle velocities. The red, green, blue and black small lines represent the ideal measurement values of the used gyroscope. The projection of the real measured values of the gyroscope present the resulting angle velocities around the single rotation axis.

The most important aspect of the curved calibration line is the continuity which is guaranteed in all points. By the continuity of the calibration curve the angle velocity can be reproduced on a resulting change of the angle. This makes an explicit determination possible.

The slope of the curved calibration line of the used gyroscope can't be described as "simple" mathematical function and is specific for each sensor. Due to this the curve is characterized as mapping function with a calibration table (lookup table).

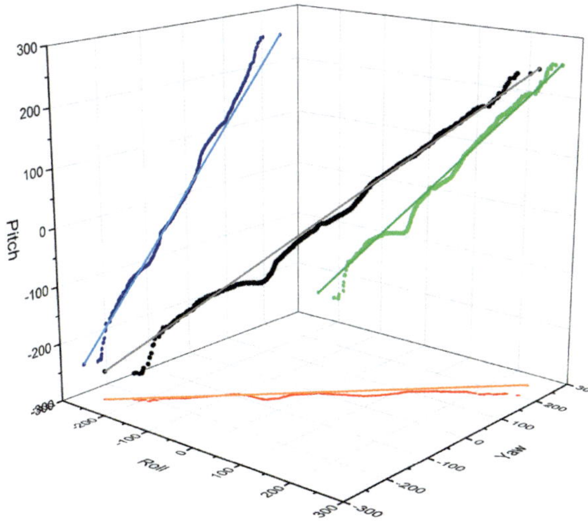

Fig. 3 *Curved line* of the calibrated gyroscope with the desired values from the data sheet

The curved line describes the correlation of the gyroscope to the accelerometer and the magnetic field sensor (ACC + MAG). This line is based on the conversion of the ACC + MAG and the transfer of these values in an angle velocity. Afterwards the calculated angle velocity of the ACC + MAG is aligned with the values of the gyroscope. By using this the calibrated value of the gyroscope is determined from the angle velocity with the sensor value of the gyroscope.

2.2 Calibration Algorithm with the Three 3D Sensors

In the last section the whole calibration algorithm is presented Köppe et al. (2013). The flow chart of the method is shown in Fig. 4.

In the first step the data of the single sensors (accelerometer, magnetic field sensor, and gyroscope) is filtered by a linear Kalman filter to minimize the sensor noise rate (SNR). Then the calibration follows as described in the sections before by the approximation of an ellipsoid. After this step a low pass filter is applied to eliminate fast movements which can be detected as an error in the calibration process. When these errors are eliminated the accelerometer and magnetic field sensor are aligned on the earth gravitation and the earth's magnetic field. At the end the calibrated values of the accelerometer and magnetic field sensor are combined with the sensor values of the gyroscope to calibrate this sensor.

As result we have a calibration method which works as permanent background algorithm to eliminate external influences. Changes of the local magnetic field are cleared and the errors which occur due to the drift properties of the sensors are eliminated. This advantage is used in combination with the definition of the measurement uncertainty in the following sections.

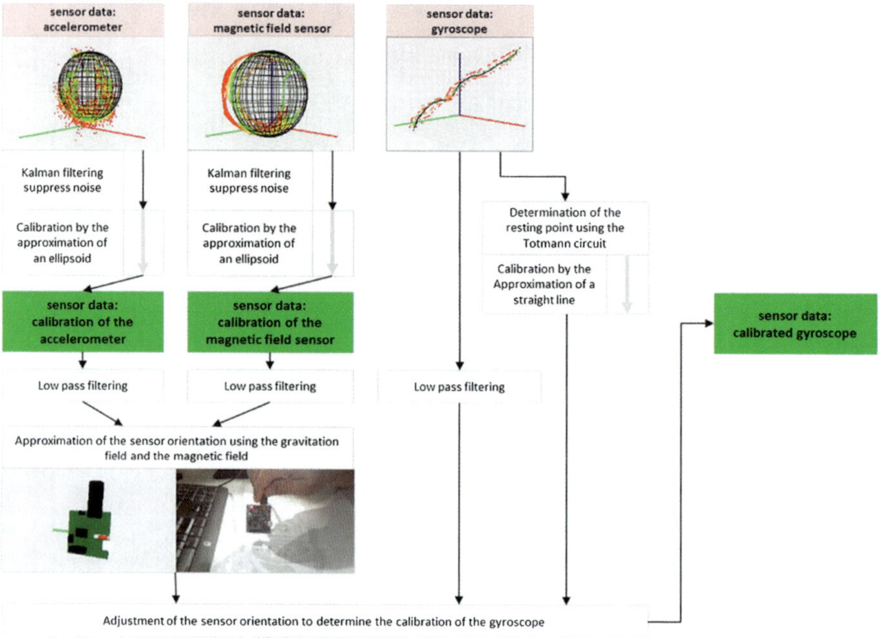

Fig. 4 Scheme of the 3D-self calibration method with all three sensors

3 Determination of the Measurement Uncertainty Using the Calibration Method

The measurement uncertainty is a parameter, associated with the result of a measurement that characterizes the dispersion of the values that could reasonably be attributed to the measurand (VIM 1993). Mostly the measurement uncertainty U is given for a stated probability of overlap of 95 %.

3.1 Theoretical Determination of the Measurement Uncertainty

At the beginning it is important to clearly define the quantity which is the subject of measurand. It is a difference if the current value of a physical quantity is measured or its mean value which is calculated from a set of values sampled over a defined time period.

Using the guideline for the determination of the measurement uncertainty (GUM 1995) then the mathematical model of a measurement is the basis for the determination of the uncertainty. It describes the functional connection between the measurement parameter Y and the input quantity X_i including all corrections with the equation $Y = f(X_i)$.

In the next step it is necessary to define the term "input value". Input values are a quantity which is measured or determined in another way to get the measurement value of the measurement parameter. They need to be considered in the mathematical model.

Furthermore there are two types of uncertainty contributions given in GUM (1995) which depend on the method of component estimation. Type A uncertainty contributions are determined by repeated measurements. Type B uncertainty contributions are estimated by other methods mainly by estimation of probability density functions and subsequent calculation of standard uncertainty (of the component). In the case of a Type B contribution known information is used, for example the tolerance of a testing equipment given by the manufacturer or the information given in standards, calibration data or from former similar input values.

The uncertainty of the value of the output quantity $Y = f(X_i)$ is a combined uncertainty which mainly consists of both types of uncertainty contributions (A and B).

In the present case repetitive calibrations and corrections are carried out using the (constant) local gravitational acceleration. An additional considerable systematic deviation is not regarded in this work and consequently any uncertainty contribution of type B is neglected. This work focuses on repetitive measurement tests, so the uncertainty evaluation is restricted to type A uncertainty contributions.

3.2 Measurement Uncertainty for the Calibration Algorithm

On the one hand for the determination of the measurement uncertainty of the accelerometer and the magnetic field sensor the method A is used, because both sensors are based on the regional condition parameters using the earth gravitation field and the earth's magnetic field. On the other hand for the gyroscope method B is used, because the determination of the measurement uncertainty for the gyroscope is based on a testing equipment (the calibrated accelerometer and magnetic field sensor). In the following Fig. 5 the influencing parameters on the input quantities are presented. The influencing parameters are the environmental conditions, the sensor, the analog to digital conversion and the calibration. The majority of the influencing parameters is normally distributed (noise, digitalization), other factors like drift, offset or linearity must be described by other estimated probability distributions obtained from prior information which correspond to type B uncertainties. These influencing parameters are corrected by the algorithms. The algorithm can be reduced with the described calibration of the sensors.

For the better illustration of the measurement uncertainty of the necessary localization sensors the measured variable, calibrated variable and sensitivity contributions for the three variable 3D sensors (ACC, MAG, and GYRO) is given in Fig. 6. The measurement uncertainty is illustrated for the eight uncertainty contributions type 1 up to type 8. Thereby the type 1–6 is available for all sensors

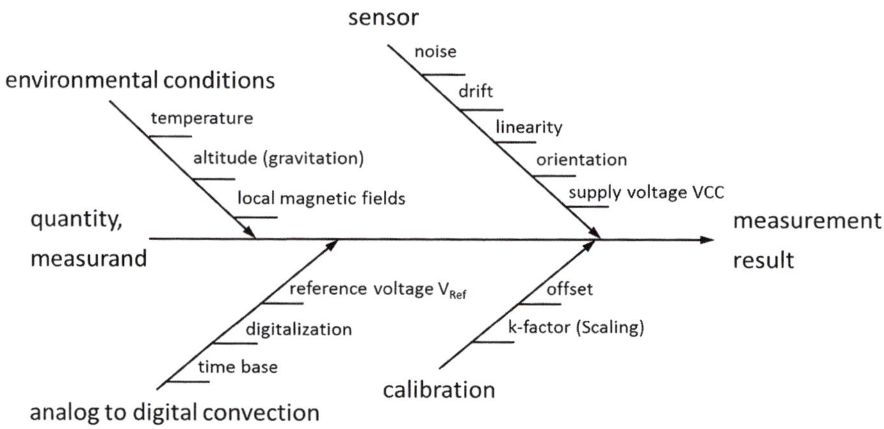

Fig. 5 Central influencing parameters and input values (quantity, measurand) for the measurement result of the 3D Sensor

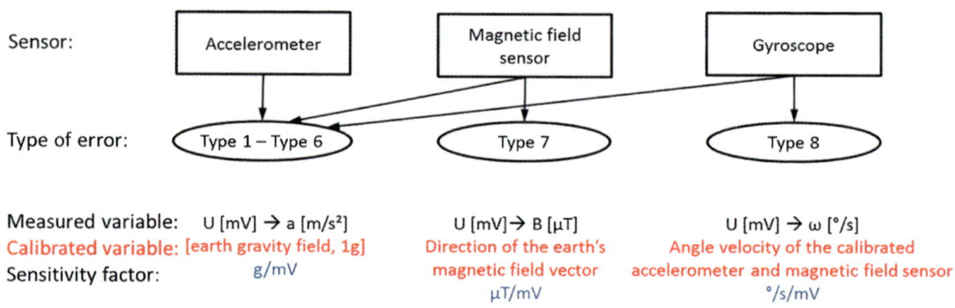

Fig. 6 Demonstration of the type of errors and the influencing variables of the three used sensors

and the type 7 is only appropriate for the magnetic field sensor as well as the type 8 is only appropriate for the gyroscope.

Furthermore the different types of uncertainty factors of the measurement uncertainty of the calibration algorithm are presented in the following.

Type 1: Resolution of the digitalization (What is the real value?)
Type 2: Sensor noise; distribution and the amplitude of the noise
Type 3: Scaling error; same sensitivity, linearity and offset
Type 4: Temperature drift; Influenced by type 1 till type 3
Type 5: Voltage fluctuation, supply voltage;
Type 6: Voltage fluctuation, reference voltage; Influence of the digitalization of the sensor values (type 1)
Type 7: Fluctuation, local magnetic fields; Interpretation of influenced sensor values by external noise fields
Type 8: Sensor drift; sensor based error factors, drift

All these uncertainty factors shortly presented above are illustrated in Fig. 7.

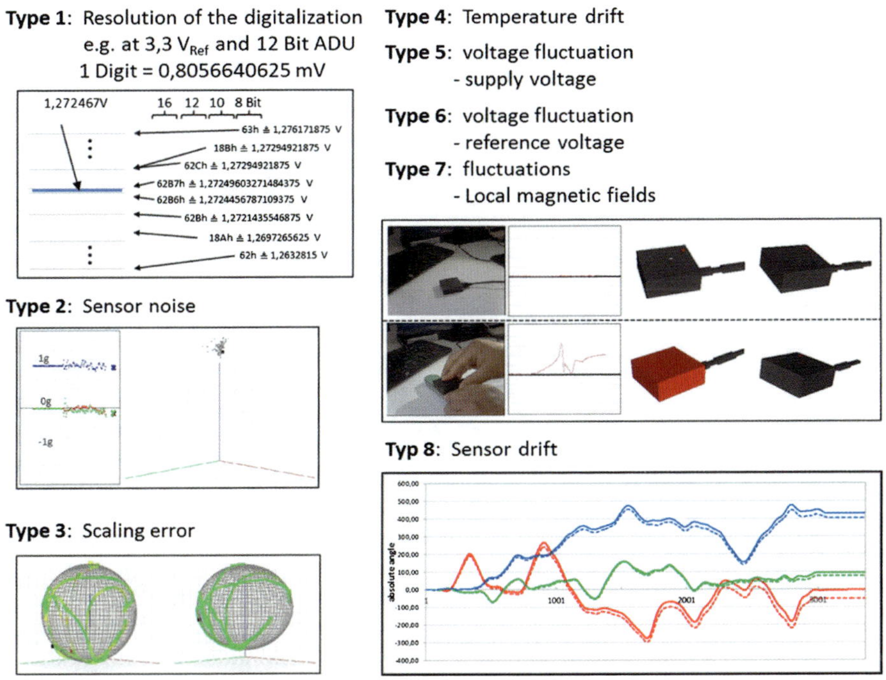

Fig. 7 Illustrated types of the uncertainty parameters of the measurement uncertainty of the three sensors (calibration algorithm)

4 Results

In this section the results for the elimination of errors by the calibration algorithm are presented. Furthermore the measurement uncertainty is reduced due to the elimination of the errors.

The influence of the calibration algorithm is shown as example at the measurement results of the gyroscope in Fig. 8. All three diagrams show the absolute angle versus the number of considered samples. The continuous line in each color shows the measured data of the gyroscope (G). In the first case (a) the uncorrected data is presented. In the second and third case (b) and (c) the corresponding corrected data (drift corrected and calibrated) is shown. The second line (the dashed line in each color) refers to the calculated calibrated data of the accelerometer and magnetic field sensor (short: MA), which does not change in all three diagrams. These values are calculated from the measured acceleration and the magnetic field which is not comparable to the absolute angle. In the top diagram (Fig. 8a) the uncalibrated measured data of the gyroscope is illustrated. A difference between the calculated and the measured data can be observed. The absolute error of the angle between the calculated and measured data curve is after 33 s (3,300 samples)

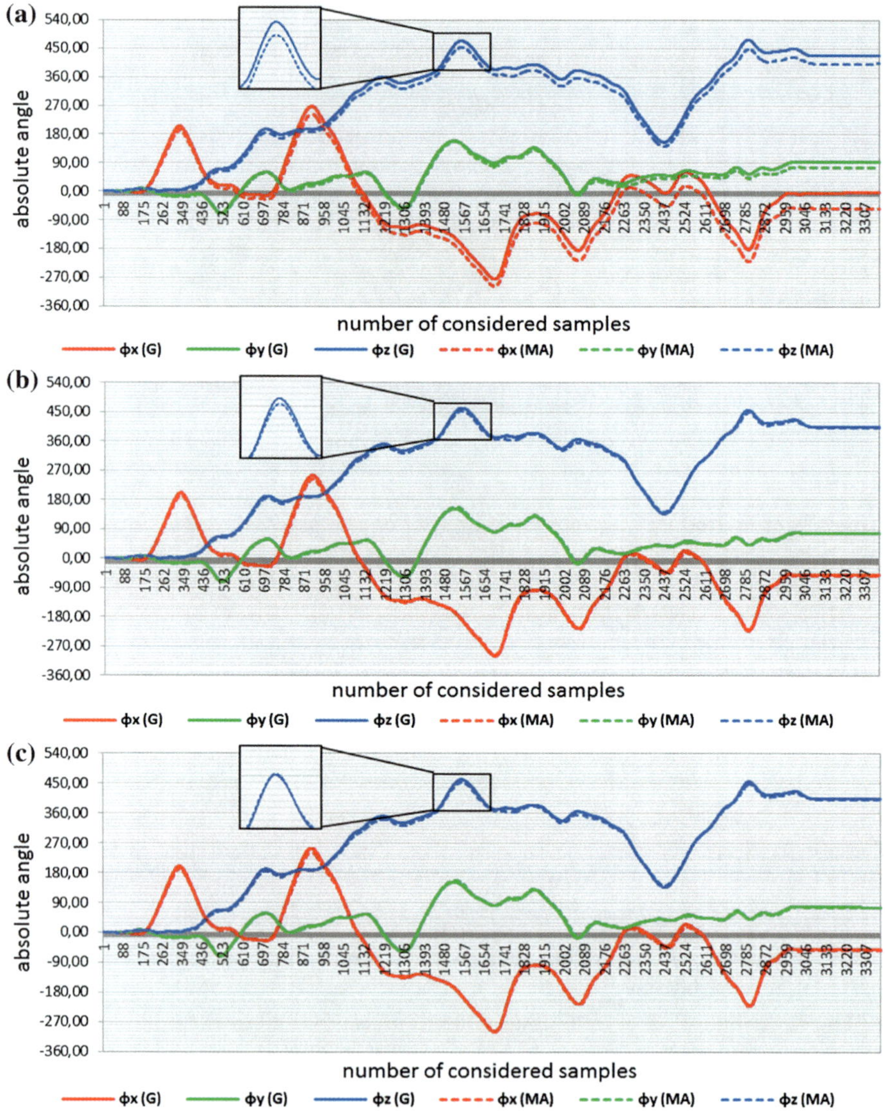

Fig. 8 Reduction of the uncertainty parameters of the gyroscope by the comparison of the data of the gyroscope and the accelerometer plus magnetic field sensor (MA): **a** uncalibrated data, **b** drift corrected data and **c** drift corrected and calibrated data

on the φx-axis −49,88°, the φy-axis −16,68° and on the φz-axis −26,94°. A peak of the curve is zoomed out in all three diagrams. This error needs to be minimized which can be done by a drift correction of the gyroscope data. The results are presented in the middle diagram (Fig. 8b). The error of the angle after the drift correction after 33 s is on the φx-axis −2,21°, the φy-axis −0,38° and on the φz-axis

−1,96°. Furthermore, the calibration of the gyroscope is performed and the result is shown in Fig. 8c. The error of the angle after 33 s is on the φx-axis 3,77°, the φy-axis −0,92° and on the φz-axis −0,80°. The deviation between the data of the MAG and the GYRO is reduced. The total error of the calibration procedure is reduced on the φx-axis from 50,89° to 16,64° and at the end to 13,10°. For the other axes the following values are given: for the φy-axis from 21,05°, 14,45° to 12,92° and for the φz-axis from 33,82°, 17,44° to 9,92°. However, the measurement uncertainties respectively the uncertainty factors type 1–6 and type 8 are reduced.

By the use of the continuous calibration algorithm presented in Sect. 2 the uncertainty factors described in Sect. 3 are reduced respectively eliminated.

Exemplary the evaluation of the self-calibration-method under consideration of the standard deviation is carried out in 126 calibration experiments. Therefore small movements were done by hand with the BodyGuard-System. These results are shown in Table 1. Looking at the three different axes of each sensor it can be seen that the deformation is around 1.0. In the table the values of the deformation are standardized and they correlate with the radii of the axes of the ellipsoid given in Sect. 2.1 for the different sensors. These values, shown in Table 1, illustrate the scaling error presented in Fig. 7 Type 3. The rest position correlates with the center point of the mentioned ellipsoids. The offset is the displacement of the center point in the different axes of the particular sensor.

For the gyroscope explained in Sect. 2.1 the deformation correlates with the slope of the line. Additional the average slope is given.

Due to the fact that the accuracy of the position determination depends on the measurement uncertainty of the measurement results of the sensors, at least the maximum standard deviation of the three sensors is shown in Table 2. All standard

Table 1 Calibration experiments with the three sensors for the evaluation of the calibration procedure

Accelerometer		
Axis	Offset of the rest position	Deformation
a_x	−15.95 mg	0.969
a_y	−33.90 mg	1.042
a_z	41.34 mg	0.961
Magnetic field sensor		
Axis	Offset of the rest position	Deformation
m_x	−81.15 mGauss	0.969
m_y	116.12 mGauss	1.042
m_z	60.27 mGauss	0.961
Gyrocope		
Axis	Average slope	Deformation
g_x	1.67 °/s	0.98
g_y	0.51 °/s	1.01
g_z	0.41 °/s	0.97

Table 2 The relation of the standard deviation of each sensor between the uncalibrated and the calibrated data of 126 experiments

Standard deviation			
Axis	Uncalibrated	Calibrated	Noise
Accelerometer Sensor			
a_x (rest position)	4.4	2.25	3.6
a_x (1 g)	6.9	2.35	
a_y (rest position)	2.9	2.27	4.5
a_y (1 g)	5.1	2.51	
a_z (rest position)	2.9	2.43	4.8
a_z (1 g)	4.6	2.38	
Magnetic field Sensor			
m_x (rest position)	1.9	0.46	4.3
m_x (local)	3.2	0.11	
m_y (rest position)	2.1	0.08	4.0
m_y (local)	2.2	0.22	
m_z (rest position)	1.3	0.22	4.1
m_z (local)	1.8	0.07	
Gyroscope			
g_x (rest position)	0.1	0.14	1.07
g_x (digit in °/s)	0.8	0.01	
g_y (rest position)	0.1	0.11	0.2
g_y (digit in °/s)	0.7	0.02	
g_z (rest position)	0.1	0.09	0.1
g_z (digit in °/s)	0.6	0.01	

deviations of the calibration values are below the noise level of each single sensor. In the table it can be seen that the calibration procedure improves the standard deviation values for all sensors.

For the accelerometer the improvement is by a factor of 2, the magnetic field sensor the improvement is by a factor of 5 and for the gyroscope it is an improvement by a factor of 8. These factors need to be validated with other high quality sensors for example fiber optic sensors and so on.

5 Conclusion

In this paper a 3D-self-calibration-method is presented for the indoor localization. This procedure was developed because of the imprecise data of low cost sensors given in manufacturer's data sheets. After the development of this calibration method the "evaluation" of this procedure with a given standard was important.

Due to this the measurement uncertainty (GUM) is determined and explained in reference to the sensors for the localization and the developed calibration method. The uncertainty parameters given in Fig. 5 of this work are determined and explained for the used sensors. At the end calibration experiments are presented to underline the improvement of the sensor data by the 3D-self-calibration-method. Furthermore the measurement uncertainty is reduced by the calibration procedure which can be seen in the results of the calibration procedure. The different types of uncertainty of the different sensors could be reduced which is exemplary presented with the drift behavior.

References

DLR (2013) German areaspace center, sensor fusion for indoor navigation, http://www.kn-s.dlr.de/indoornav/

Google (2014) Google Projekt Tango, Abteilung für fortgeschrittene Technologien (ATAP). https://www.google.com/atap/projecttango/

GUM (1995) Guide to the expression of uncertainty in measurement, 1st edn, 1993, corrected and reprinted 1995, International Organization for Standardization, Geneva. www.bipm.org/utils/common/documents/jcgm/JCGM_100_2008_E.pdf

Köppe E (2012) Radio-based multi-sensor system for person tracking and indoor positioning, WPNC 12:180–186

Köppe E, Augustin D, Liers A, Schiller J (2012) Automatic 3D calibration for a multi-sensor system IPIN 2012, Nov 2012. doi:10.1109/IPIN.2012.6418870

Köppe E, Augustin D, Liers A, Schiller J (2013) Enhancement of the automatic 3D calibration for a multi-sensor system IPIN 2013, pp. 234–236, Oct 2013

Köppe E, Augustin D, Liers A, Schiller J (2014) Self-calibration-method for an inertial navigation system with three 3D sensors IEEE ISISS 2014. http://ieeexplore.ieee.org/xpl/articleDetails.jsp?arnumber=6782522, doi:10.1109/ISISS.2014.6782522, Feb 2014

Köppe E (2014) Lokalisierung sich bewegender Objekte innerhalb und außerhalb von Gebäuden (German), Dissertation, FU Berlin, Department of Mathematics and Computer Science. http://www.diss.fu-berlin.de/diss/receive/FUDISS_thesis_000000097278, Aug 2014

Microsoft (2014) Microsoft Research, Mobile indoor localization. http://www.research.microsoft.com/en-us/projects/indoorloc

Schmid J (2012) Ad-Hoc Personenlokalisierung in Drahtlosen Sensornetzwerken (German), Dissertation, Karlsruhe Instituts of Technology (KIT). http://digbib.ubka.uni-karlsruhe.de/volltexte/1000030410

Titterton DH, Weston JL (2004) Strapdown inertial navigation technology, 2nd Edition (IEE Radar, Sonar, Navigation and Avionics, Series 17) Institution of Engineering and Technology

VIM (1993) International vocabulary of basic and general terms in metrology, 2nd edn. International Organization for Standardization, Geneva

Part III
Spatial-Temporal Data Processing and Analysis

Feature Selection in Conditional Random Fields for Map Matching of GPS Trajectories

Jian Yang and Liqiu Meng

Abstract Map matching of the GPS trajectory serves the purpose of recovering the original route on a road network from a sequence of noisy GPS observations. It is a fundamental technique to many Location Based Services. However, map matching of a low sampling rate on urban road network is still a challenging task. In this paper, the characteristics of Conditional Random Fields with regard to inducing many contextual features and feature selection are explored for the map matching of the GPS trajectories at a low sampling rate. Experiments on a taxi trajectory dataset show that our method may achieve competitive results along with the success of reducing model complexity for computation-limited applications.

Keywords Map matching · GPS trajectory · Conditional random fields · Feature selection

1 Introduction

Map matching of GPS trajectory serves the purpose of recovering the original route on a road network from a sequence of GPS observations. It is a fundamental technique for many Location Based Services (LBS) as it brings added value to the raw GPS data and has the potential to distill more reliable knowledge about routing on road networks. However, the GPS observations are often noisy so that finding the nearest roads usually fails. Many research works have been dedicated to map matching of GPS trajectory with a moderate sampling rate, while map matching with a low sampling rate, namely the sampling interval greater than 120 s, is still an ongoing research topic in recent years (Hunter et al. 2013; Li et al. 2013).

Map matching is often modeled as a sequence labeling problem. The Hidden Markov Model (HMM) and its variants have been intensively explored in previous

J. Yang (✉) · L. Meng
Lehrstuhl für Kartographie, Technische Universität München, 80333 Munich, Germany
e-mail: jian.yang@tum.de

G. Gartner and H. Huang (eds.), *Progress in Location-Based Services 2014*,
Lecture Notes in Geoinformation and Cartography,
DOI 10.1007/978-3-319-11879-6_9

attempts (Hummel 2006; Krumm et al. 2007; Lou et al. 2009; Newson and Krumm 2009; Yuan et al. 2010). Being constrained by the strict statistical assumptions, however, these generative models fail to capture the non-independent characteristics from sparse GPS observations and therefore result in poor performance. This gives rise to Conditional Random Fields (CRFs) (Lafferty et al. 2001), another probabilistic model for labeling sequential data that allows to use many non-independent and overlapped features drawn from observations to improve the matching accuracy. However, the CRFs requires intensive computation which could prohibit computation-limited applications such as the map matching on mobile devices. This constraint stimulates the need to select the most relevant feature subset in the CRFs, thus reduce the model complexity in terms of the number of features.

In this paper, we attempt to construct a compact CRFs for map matching through feature selection. More specifically, we first induce rich features to CRFs for map matching, and then train the CRFs with ℓ_1 regularization to yield a sparse model (many features are assigned to zero weights). To verify the effectiveness of feature selection, we perform an experiment on a sample dataset derived from Taxi Floating Car Data (FCD) in Shanghai, China. Following contributions could be highlighted in our work:

1. We explore the further use of the CRFs for map matching to yield a sparse model with a higher matching accuracy via feature selection. Experiment shows that 50 % feature reduction and 10 % accuracy improvement can be achieved compared to a common model.
2. As we induce features from most cited literatures, the result of feature selection can serve as an experimental review of previous modeling effort and provide guidance for designing simple model for map matching.
3. The learned weights of selected features also reflect road usage pattern in the study area.

2 Map Matching of GPS Trajectory

Map matching requires both GPS observations and a road network. The basic attributes of the observations collected by positioning sensors include latitude, longitude and timestamp, while extra information such as instant speed, acceleration, heading direction etc. can also be obtained from the sensors. Due to the inaccuracy embedded in both observations and the road network, the matching of the nearest road matching often fails and it is therefore necessary to develop map matching methods. The challenge of this task is two-folded: (1) Observations are often noisy due to inaccurate GPS sensor or poor positioning conditions, e.g. low-speed maneuvers of vehicle in traffic, passage through urban canyon and tunnels. These facts make map matching problematic in a dense road network. (2) A low sampling rate which aims to reduce the communication cost or data storage causes

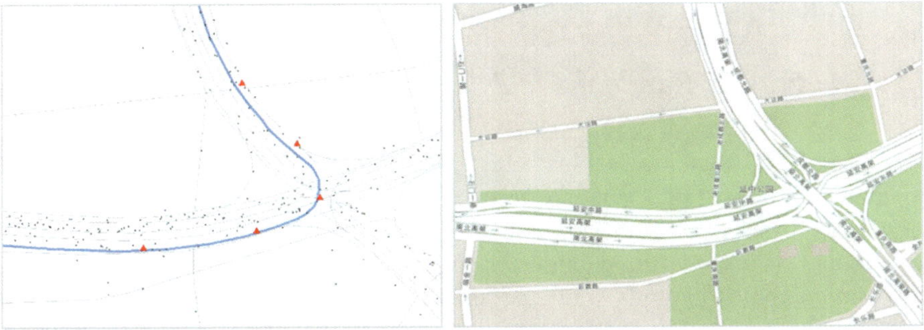

Fig. 1 An example of GPS trajectory section in a dense road network. *Left* GPS observations depicted as *red* triangle along the true path in *blue* in the vector representation. The sampling interval is 10 s. *Right* Map view of roads in that area

information loss between neighboring observations, making the route recovery extremely difficult as a huge number of feasible paths can be found on road network. An example of GPS trajectory is illustrated in Fig. 1.

Map matching has invoked a growing interest in the past years for its importance in LBS applications. A comprehensive literature survey was done in (Quddus et al. 2007), in which map matching method is categorized into four groups: geometric, topological, probabilistic and other advanced techniques. Among these approaches, the The Hidden Markov Model (HMM)-based probabilistic methods are most popular because of their well-studied theoretical base and competitive performances. A HMM-based method models the probability of a sequence possible road assignments on road network for given GPS observations. The computation of the sequence probability requires a strict modeling of the observation probability and transition probability, namely probability for candidate roads for GPS observations and candidate paths in between. These two components are designed to capture the characteristics of noisy sensors and the original route from different perspectives. And matching GPS trajectories at a low sampling rate often requires richer features of observation and transition for better accuracy. However, this would cause an intractable inference problem for HMM models (Lafferty et al. 2001).

(Hunter et al. 2013) first introduced the CRFs to map matching in a real world project and achieved the best performance for a sampling interval of 60 s by building a complex model using 10 features. The CRFs shares with the HMM a similar factorization of probability computation, but is more flexible in using non-independent features. The success in practice and the flexible nature of CRFs has motivated us to incorporate complex features by leveraging existing modeling efforts in HMM-based methods. However, using a large number of features in the CRFs could cause over-fitting and increase computational expenses. To overcome this potential drawback, we investigate a ℓ_1 regularized CRFs with the aim to set up a sparse model.

3 Map Matching with Conditional Random Fields

3.1 Conditional Random Fields

The Conditional Random Fields (CRFs) is an undirected graphical model used to compute probability of a possible label sequence conditioned on the observation sequence (Lafferty et al. 2001). The CRFs represents the conditional probability as the product of potential functions over cliques in the graph. These potential functions are computed in terms of feature functions of random variables in observation and label sequence. Let $Y = \{y_1, y_2, \ldots, y_T\}$ and $X = \{x_1, x_2, \ldots, x_T\}$ denote the label sequence and observation sequence. A CRFs formulates conditional probability of Y given X as:

$$P(Y|X) = \frac{1}{Z}\prod_q \exp\left(\sum_k \omega_k f_k(Y_q,\ X_q)\right)$$

where $\{f_k\}$ are the feature functions on any subset of the random variables in the sequence $Y_q \subset Y,\ X_q \subset X$ (note that $Y_q \cup X_q$ form the cliques in the graph), $\{\omega_k\}$ are the trained weights for each feature function, and Z is a input-dependent normalization term over all possible state sequence:

$$Z = \sum_Y \prod_q \exp\left(\sum_k \omega_k f_k(Y_q, X_q)\right)$$

3.2 A CRFs Framework for Labeling GPS Trajectory

To apply the CRFs to map matching, we first define random variables to model the observation and label sequence. Let $X = \{x_1, \ldots, x_N\}$ be GPS observation sequence, $Y = \{y_1, \ldots, y_{2N-1}\}$ be the label sequence, N be the length of the observation sequence and $t = 1\ldots N$ be the position index in the sequence. We give the definition as follows:

- $x_t \in X$ is a variable representing GPS observation.
- $y_{2t-1} \in Y, t = 1\ldots N$ is a random variable over point states $R^t = \{r_i^t\}, i \in N_{2t-1}$ of observation x_t, where R^t is a finite set of nearby roads of x_t within a predefined distance.
- $y_{2t} \in Y, t = 1 \ldots N-1$ is a random variable over path states $P^{2t} = \{p_i^{2t}\}$, $j \in N_{2t}$, where P^{2t} contains all the feasible paths between road r_i^t and road r_i^{t+1}. And P^{2t} is also a finite set since vehicle can only travel a limited distance in a road network in specific time duration with speed limits.

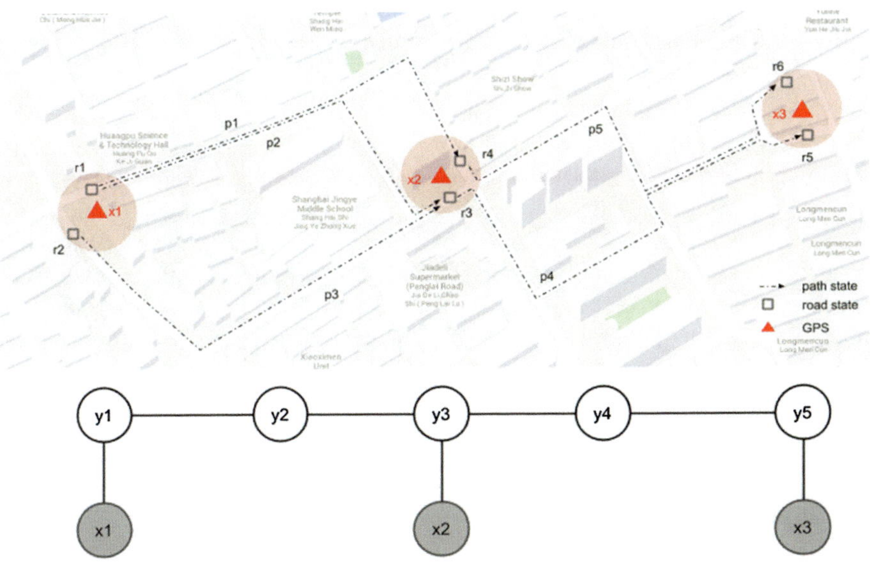

Fig. 2 A chain-structured CRFs for 3 GPS observations. The map on *top* illustrates the simplified situation of identifying road states and path states given GPS observations in the road network. This requires 5 random variables, $y_1 : \{r_1, r_2\}$, $y_2 : \{p_1, p_2, p_3\}$, $y_3 : \{r_3, r_4\}$, $y_4 : \{p_4, p_5\}$, $y_5 : \{r_5, r_6\}$, to build the CRFs for map matching. Thus, nodes y_1, y_3, y_5 linking with observations (*black circles*) are point nodes while nodes y_2, y_4 are path nodes

Take abovementioned variables as the nodes, in which we call $\{x_t\}$ observation node, $\{y_{2t-1}\}$ point node and $\{y_{2t}\}$ path node. Then, we add edges between observation nodes and point node at each position t, while linking point nodes and path nodes sequentially. To be more concrete, we give a simplified example of chain structured CRFs for 3 GPS observations on road network in Fig. 2. Note that applying different features to the model could result in a different topology between variable nodes and observation nodes.

Then, a chain-structured CRFs for map matching is formulated as:

$$P(Y|X) = \frac{1}{Z}\prod_{t=1}^{N} \exp\left(\sum_{k=1}^{K} \omega_k f_k(y_{2t-1}, x_t) + \sum_{s=1}^{S} \mu_s g_s(y_{2t}, y_{2t-1}, y_{2t+1}, X)\right) \quad (1)$$

where $\{f_k\}$ are feature functions defined on point nodes while $\{g_s\}$ are feature functions defined on path nodes (note that $g_s = 0$ when $t = N$), $\{\omega_k\}$ and $\{\mu_s\}$ are weights of the feature functions. These feature functions are designed to capture the characteristics of the actual label of point and path respectfully which we will discuss in detail in Sect. 4. And Z is given as:

$$Z = \sum_{y \in Y} \prod_{t=1}^{N} \exp\left(\sum_{k=1}^{K} \omega_k f_k(y_{2t-1}, x_t) + \sum_{s=1}^{S} \mu_s g_s(y_{2t}, y_{2t-1}, y_{2t+1}, X)\right) \quad (2)$$

The rationale of using path node to explicitly model transition between neighboring observations is that it allows the model to evaluate more than one path between two road states. This may avoid an early elimination of truth path state as most HMM-based methods are forced to use only one path, e.g. the shortest path in most cases. The modification is crucial especially for a low sampling rate of GPS trajectory which may lead to identification of many feasible paths. And our model differs from that by (Hunter et al. 2013) in the way that we encode the transition on the path node rather than on edges, which helps reduce the space complexity in later implementation.

3.3 Inference and Training on CRFs

Map matching can be casted as an inference on CRFs, which is to find the state sequence with the maximal probability conditioned on the observation sequences. For a general structured CRFs, the inference could become computationally intractable because the increasing length of the trajectory will exponentially enlarge the resulting state space. However, an exact solution can be obtained using dynamic programming algorithms such as Vite-bi on linear chain structure (Sutton 2012).

The inference requires learned weights of the feature functions, which can be estimated by training CRFs with labeled data, namely GPS observation sequences are labeled against actual road sequences (also the road sequences in between). A common training scheme is to estimate the weights of the feature functions by maximizing the log likelihood function $\log(P(Y|X; \omega, \mu))$, which yields

$$\begin{aligned} \ell(\omega, \mu) &= \log(P(Y|X; \omega, \mu)) \\ &= \sum_{t=1}^{N}\left(\sum_{k=1}^{K} \omega_k f_k(y_{2t-1}, x_t) + \sum_{s=1}^{S} \mu_s g_s(y_{2t}, y_{2t-1}, y_{2t+1}, X)\right) \quad (3) \\ &\quad - \log(Z) \end{aligned}$$

Since feature functions for point node and path node can be equally treated in the optimization, we rewrite the objective function for brevity as follows

$$\ell(\theta) = \sum_{t=1}^{N}\left(\sum_{m=1}^{K+S} \theta_m q_m\right) - \log(Z) \quad (4)$$

where

$$(\theta_m) = (\omega_1, \omega_2, \ldots; \ \mu_1, \mu_2, \ldots)$$
$$(q_m) = (f_1(\cdot), f_2(\cdot), \ldots; \ g_1(\cdot), g_2(\cdot), \ldots)$$

This yields a convex and differentiable objective function for which we can use unconstrained optimization method to find the global optimal solution. More specifically, a quasi-Newton method, BFGS, is used, which has been found successful in terms of efficiency for solving this objective function (Sha et al. 2003).

4 Feature Selection in CRFs with ℓ_1 Regularization

Often, to improve the classification result of the CRFs, more features should be used. However, this leads to a dilemma that using more features also increases the risk of over-fitting. Therefore, it has been a long-term endeavor in machine learning community to study feature selection with the aim to find the most relevant feature subset to build a compact and interpretable model (Ng 1998). This involves two tasks, feature induction and feature selection. We discuss them in the following sections in the context of map matching.

4.1 Feature Induction and Parameter Tying

In our chain structured model, two types of feature functions are used, namely point features and path features. Both features can be designed in an either manual or automated fashion to capture the characteristics of truth states. We employ both strategies for the feature induction.

Hand-crafted features are extracted from HMM-based map matching methods (Goh et al. 2012; Hummel 2006; Krumm et al. 2007; Lou et al. 2009; Newson and Krumm 2009; Yuan et al. 2010). As HMM shares with the CRFs a similar structure, it is straightforward to derive point feature and path feature from the emission probability and the transition probability. Note that most of these probabilities follow the assumption of Gaussian distribution. Since the CRFs uses exponential parameterization, only the informative power terms in the formulation of HMM-based methods are needed. The detailed formulation is given in Sect. 5.

Another observation for map matching is that routing on road network is a dynamic process. Driver behaviors could vary a lot in different spatiotemporal contexts. For example, paths along main roads could be avoided to dodge heavy traffic in rush hours while they are taken during the night. In order to feed the model with this kind of contextual information, we employ two feature templates coding road class and temporal information for point and path nodes:

$$f = \mathbf{I}(y_{2t-1} = r_i^t)\, \mathbf{I}(r_i^t = \text{class}_u)\, h(y_{2t-1}, X) \tag{5}$$

$$g = \mathbf{I}(y_{2t} = p_j^{2t})\, \mathbf{I}(t \text{ in period}_v)\, h(y_{2t}, X) \tag{6}$$

where **I** is indicator function, it yields 1 when the given condition holds, 0 otherwise, *h(y, X)* could be any feature functions revealing contextual characteristics of the ground truths (e.g., varying travel speed in different periods during the day), $\{\text{class}_u\}$ and $\{\text{period}_v\}$ are sets of road classes and time periods in hour during the day. The road class information is extracted from OpenStreetMap (OSM) and timespans are divided using predefine time interval.

Feature template could result in a large number of features. For example, *M* feature functions with *N* would need *MN* parameters. This imposes a heavy computational load for the later training as *N* is usually very large for the low sampling rate trajectory. However, the CRFs for map matching serves as a structured classifier to perform binary classification of truth/false state for both point node and path node. And there is no need to assign a unique parameter for individual point states and path states at each position in the chain. Therefore, only *M* parameters are needed in the model.

4.2 Training CRFs with ℓ_1 Regularization

To achieve a better classification performance in map matching, a large number of features are used in the CRFs. This yields a lower error rate on training data while raising the risk of high generalization error on test data. A common technique to tackle this problem is to add a penalty term to the objective function which penalizes learning large weights of feature functions in training. In this section, we discuss two kinds of regularization techniques, ℓ_2 regularization and ℓ_1 regularization, and explain how to perform the feature selection with the latter one.

1. ℓ_2 Regularization

ℓ_2 regularized CRFs adds a negative quadratic term to the objective, which tends to keep all the weights small enough but non-zero in training. This yields to solve:

$$\max_{\theta} \ell(\theta) - \lambda_2 (\sum_m \theta_m) \tag{7}$$

Here $\lambda_2 \geq 0$ is a hyper parameter that controls the amount of the penalty: the larger the value of λ_2, the greater the amount of penalty and 0 for no penalty. Since the penalty term is differentiable with respect to parameters of the model, the objective remains convex and differentiable. Therefore the optimization method used to train non-regularized CRFs can also be applied here.

2. ℓ_1 Regularization

Another regularization technique, ℓ_1 regularization, adds absolute term to the objective, which tends to reduce the weights to exactly zero in training. This yields to solve:

$$\max_{\theta} \ell(\theta) - \lambda_2 \sum_{m} |\theta_m| \tag{8}$$

where $\lambda_1 \geq 0$ again is used to tune the amount of penalty. The objective also remains convex while become non-differentiable at $\theta_m = 0$, which requires extra treatment to solve this optimization problem.

Having the advantage of producing a sparse model (having many parameter set to 0), optimizing ℓ_1 regularization has invoked a lot of interest in machine learning community. A variety of optimization methods are proposed to solve the problem. Since the convexity of ℓ_1-regularized objective ensures the finding of a unique optimal solution, those methods can be distinguished by how they handle non-differentiability of the objective function. Therefore, we mainly consider the efficiency in terms of running time while choosing optimization algorithms. Some comprehensive experimental reviews have been reported in (Schmidt et al. 2009; Schmidt 2010), which stimulated our interest in the Projected Scaled Sub-Gradient (PSS) methods for its fast convergence rate and consistent performance across different types of data set. We also find it more successful on GPS trajectory data.

Still, we have to choose the hyper parameters λ_1 and λ_2 which are difficult to determine in advance. As for λ_1, we tune the hyper parameters by evaluating the resulting error rates using a geometric sequence of decreasing from $\lambda_{\max}$ to 0, where $\lambda_{\max}$ is large enough to reduce all weights to zero. The justification of using a geometric sequence is that the target value is close to 0 and more trials are needed to approach it. And we use the same hyper parameter for ℓ_2 for comparison.

5 Experiment

We build a compact CRFs for map matching of low sample rate GPS trajectories by training the model with ℓ_1-regularization. To examine the efficiency of ℓ_1-regularization, we test our methods on a GPS trajectory sample dataset ST100 to compare its error rates of map matching with the CRFs trained with a common ℓ_2-norm. In following sections, we first introduce the sample dataset, and then describe features for map matching. In the end, the results are discussed.

5.1 Experiment Setup

The sample dataset ST100 records GPS trajectories of 70 taxis from one day across the downtown area in Shanghai, China. It involves 124 trajectories in total and 13,767 GPS observations covering an overall length of 788 km after eliminating some erroneous trajectories, e.g. extremely short trips and trips losing long distance GPS observations. Spatial distribution of the trajectories in ST100 and statistics of sample trajectories are demonstrated in Fig. 3.

As the data source doesn't provide the ground truth labels, we have to manually label them on the reference road network. We recruited 2 volunteers with driving experiences in China to trace the trajectories on the map using OSRM, a web-based interactive routing application using road data from OSM. For routing exceptions like U-turns, it requires manual post-processing individually.

In order to test the consistency of our models at different sampling rates, we degrade ST100 to three datasets with 60, 90 and 120 s interval accordingly using an even sampling strategy. For each degraded dataset, we split it into a training set and a test set with a ratio of 7:3. A portion of training set is used as hold out data to tune the hyper parameters. These settings are applied to both ℓ_2 and ℓ_1 regularization.

5.2 Features for Map Matching

Here we give details of the features used in the CRFs for all experiments. There are in total 61 features in the model. 9 of them are derived from HMM-based methods in the literatures and most of the rest are designed to reveal the road usage pattern and temporal behavior of the drivers. For brevity, we omit the dummy term $\mathbf{I}(y_{2t-1} = r_i^t)$ for point node feature and $\mathbf{I}(y_{2t} = p_m^{2t})$ for path node features. And we set time interval to 4 h for all temporal features (Table 1).

These feature data are generated from the ST100 and the OSM road network on Postgresql with PostGIS and pgRouting. The spatial extensions are used to perform spatial queries and graph search to identify road states and path states. Before the data are fed to the CRFs, we rescale the features to the range [0, 1] so as to avoid a dominant impact of some features with large values on the model.

5.3 Matching Results of Low Sampling Rate GPS Trajectory

We tested our model with two different regularization on three sets of low sampling rate GPS trajectories from sample dataset ST100 and compared the error rate of point and path separately. Though point nodes and path nodes have mutual impacts on each other in the chain structure of the CRFs, labeling path is usually more

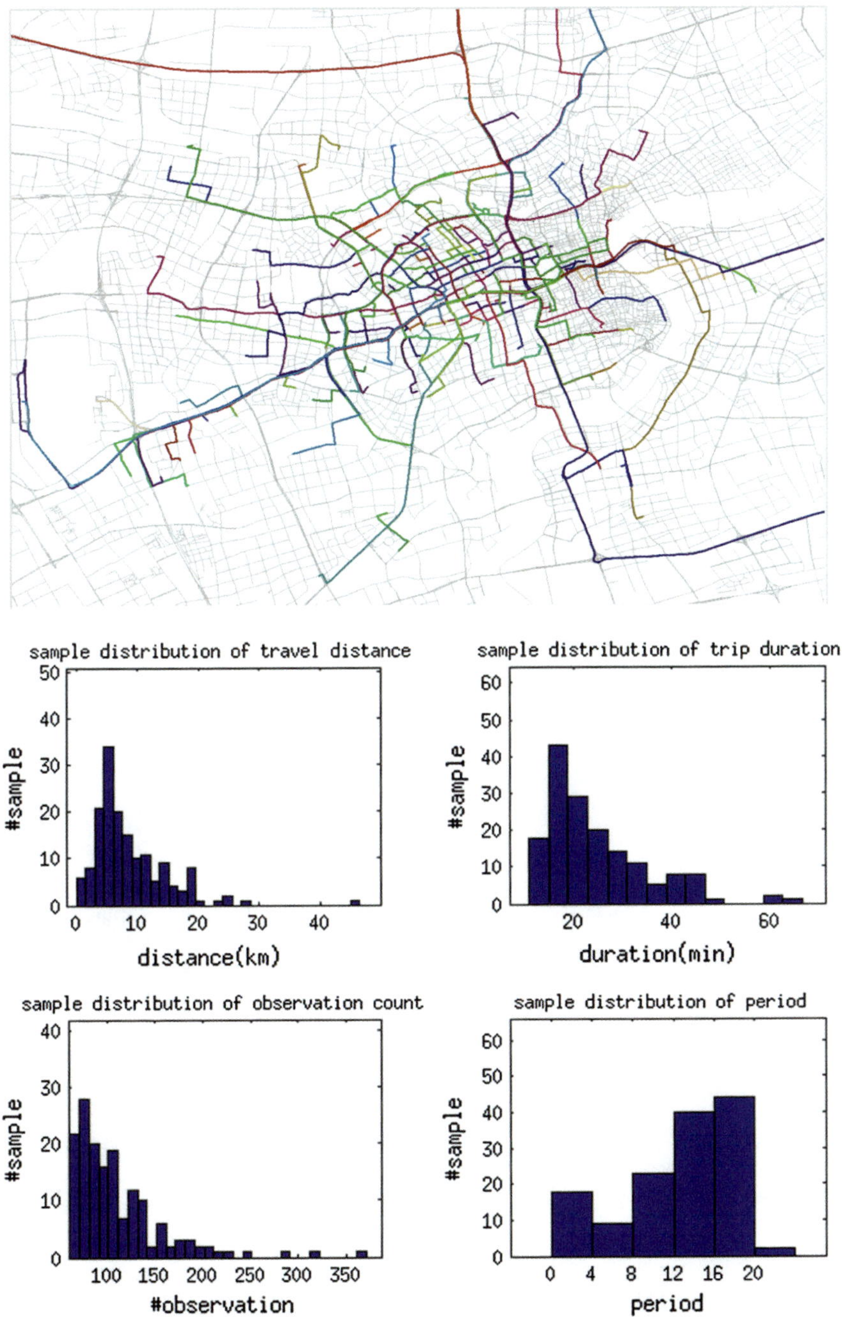

Fig. 3 (*Top*) The spatial distribution of GPS trajectories in ST100. *Bottom* The statistics of sample trajectory in ST100: travel distance *upper left*, trip duration *upper right*, observation count *bottom left* and daytime period in hour *bottom right*

Table 1 Features used in the model

Feature	Description	Node type
$\text{dist}(x_1, r_i^t)$	*GPS distance error* between GPS observation and road candidate	Point
aziumth (x_t, r_i^t)	*angular difference* between vehicle's heading direction and the road direction	Point
$(v(x_t) - v(r_i^t))/v(r_i^t)$	*speed difference* ratio between vehicle and speed limits of the road	Point
$\mathbf{I}(t \text{ in period}_v)(v(x_t) - v(r_i^t))/v(r_i^t)$	*temporal speed difference ratio*	Point
$\mathbf{I}(r_i^t = \text{class}_u)\mathbf{I}(\text{taxi in service})$	*road usage* indicate how often a certain class of roads is used when the taxi is with passenger	Point
$\mathbf{I}(r_i^t = \text{class}_u)\mathbf{I}(t = 1 \ or\ N)$	*IO feature* indicates the road usage when the taxi picks up or drops off passengers	Point
$\text{length}(p_j^{2t})$	*length* of the path	Path
$\text{t}_{\min}(p_j^{2t})$	*minimum travel time* on the path, using speed limits of the roads and the time interval between GPS observations	Path
$\bar{v}(p_j^{2t})$	*maximum average speed* is the average speed limits of roads in the path	Path
$\text{dist}(x_t, x_{t+1})/\text{length}(p_j^{2t})$	*length ratio* of distance between GPS observations to the path's length	Path
$\cos(v(p_j^{2t}), \bar{v}(p_j^{2t}))$	*cosine distance* between the speed limits of the roads in path and the overall average speed on the path	Path
$\text{length}(p_j^{2t}) - \text{dist}(x_t, x_{t+1})$	*length difference* between the path and the distance between GPS observations	Path
$\text{t}_{\min}(p_j^{2t}) - t(x_t, x_{t+1})$	*time difference* between the estimated the minimum travel time and the actual travel time	Path
$\text{classmod}(p_j^{2t})$	*road class changes* in the path	Path
$\mathbf{I}(t \text{ in period}_v)(\text{length}(p_j^{2t}) - \text{dist}(x_t, x_{t+1}))$	*temporal length difference* of the path	Path
$\mathbf{I}(\text{taxi in service})\text{classmod}(p_j^{2t})$	*temporal road class changes*	Path

difficult than labeling points. Therefore, we chose to evaluate the performance on different nodes individually.

The matching results on three sampled datasets are summarized in Table 2. It shows that the error rate increases as the sampling interval grows in which the path error rates deteriorate faster than the point error rates. The reasons that path matching is more challenging are (1) path features fail to discriminate the actual paths among the huge numbers of the path candidates; (2) the routing preference might not be consistent across the trajectories. With regard to the two regularizations, training CRFs with ℓ_1-norm managed to reduce more than half of the features

Table 2 Error rates of CRFs with different regularizers for map matching

Intervals	Regularizer	Feature number	Point error rate	Path error rate
60	ℓ_2	44	0.228	0.299
	ℓ_1	18	0.153	0.194
90	ℓ_2	43	0.235	0.304
	ℓ_1	20	0.146	0.197
120	ℓ_2	43	0.255	0.339
	ℓ_1	17	0.166	0.234

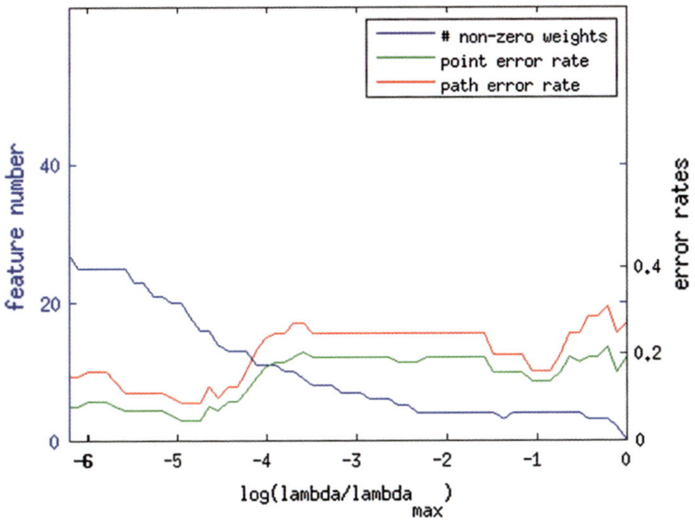

Fig. 4 The impact of tuning λ on the number of selected features (non-zero weights) and error rates

that are necessary with ℓ_2 while achieving an average of 10 % reduction on the error rates. Meanwhile, we also compare ℓ_1-regularized CRFs with *MaxLL-complex* from (Hunter et al. 2013). Both achieve only the same accuracy performance on the test data of 120 s interval. However, our method is more flexible because we give alternative choices of features, which could be helpful when desired features are not available in the data (e.g., POIs of traffic lights are not available in the test area in OSM).

We examined the effectiveness of feature selection by tuning hyper parameter on hold out dataset with 120 s sampling intervals. As shown in Fig. 4, features are gradually added to the model when decreasing λ. And the error rate is reduced as more informative features are used. The improvement stopped at some tipping point where adding more features may cause the model to over fit the training data.

In the end, we evaluate all the selected features (only those have non-zero weights in all three tests) in Fig. 5. The weights vary dramatically across the

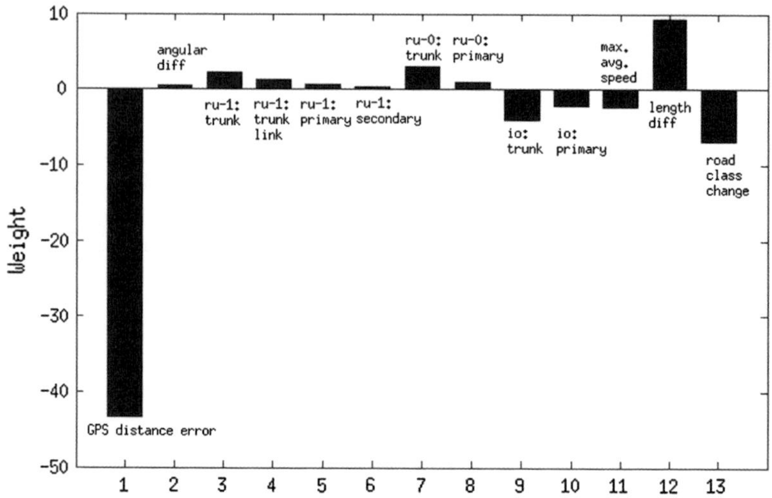

Fig. 5 The weights of selected features learned from the data

selected features, in which 1–10 are for point features and 11–13 are for path features. The weight magnitude shows the relevance degree of the feature to the map matching task, while the sign of weights indicate how the features are related. For example, *GPS distance error* getting a negative weight means that the states are more likely to be true if the its *GPS distance error* are smaller. Among all the features, *GPS distance error* (1), *length difference* (12) and *road class changes* (13) are the most relevant ones. The negative weight of *road class change* indicates that the drivers prefer to stay on the roads of the same class. The weights of feature *road usage* (3–8) show that the road usages are unbalanced between the taxis in service (ru-1) and not (ru-0). Two classes of roads are selected in the feature *IO* (9–10) with relatively large negative weights indicating that taxis may barely pick up and drop off passengers there.

6 Conclusion and Future Work

By inducing complex and non-independent features, we explored the use of CRFs for map matching GPS trajectory at a low sample rate. Rather than using a common ℓ_2-regularization, we train the CRFs with a ℓ_1-norm to yield a sparse model which requires less computation cost to perform the map matching. To verify the model, we build a sample dataset, ST100, from Shanghai Taxi FCD. Experiments on ST100 have shown the effectiveness of ℓ_1-regularization on both feature selection and matching accuracy. The result of feature selection can provide a guidance to build a compact model and meanwhile reveals to a certain extent the pattern of road usage in the urban road network of our study area.

In the future work, we intend to improve the method from following perspectives: (1) induce context-aware features to capture the spatial variance of routing decisions on urban road network for map matching; (2) study efficient training method for the CRFs with a larger feature set.

Acknowledgments We would like to thank Dr.-Ing Hongchao Fan and Prof. Chun Liu for sharing with us the Shanghai Taxi FCD dataset, and to Oliver Maksymiuk for the helpful discussion. The first author is supported by China Scholarship Council (CSC).

References

Goh C, Dauwels J, Mitrovic N (2012) Online map-matching based on Hidden Markov model for real-time traffic sensing applications. In: ITSC 12'

Hummel B (2006) Map matching for vehicle guidance. In: Drummond J, Billen R (eds) Dynamic and mobile GIS: investigating space and time. CRC Press, Florida

Hunter T, Abbeel P, Bayen A (2013) The path inference filter: model-based low-latency map matching of probe vehicle data. Algorithmic foundations of robotics X. Springer, Berlin, pp 591–607

Krumm J, Letchner J, Horvitz E (2007) Map matching with travel time constraints. In: SAE world congress

Lafferty J, McCallum A, Pereira F (2001) Conditional random fields: probabilistic models for segmenting and labeling sequence data. In: ICML 2001 pp 282–289

Li Y, Huang Q, Kerber M (2013) Large-scale joint map matching of GPS traces. In: ACM GIS'13, pp 1–10

Lou Y, Zhang C, Zheng Y, Xie X, Wang W, Huang Y (2009) Map-matching for low-sampling-rate GPS trajectories. In: ACM GIS'09. ACM Press, p 352

Newson P, Krumm J (2009) Hidden Markov map matching through noise and sparseness. In: Proceedings of the 17th ACM GIS '09. p 336

Ng AY (1998) On feature selection: learning with exponentially many irrevelant features as training examples. In: ICML'98. pp 404–412

Quddus M, Ochieng W, Noland R (2007) Current map-matching algorithms for transport applications: state-of-the art and future research directions. Transp Res Part C: Emerg Technol 15(5):312–328

Schmidt M (2010) Graphical model structure learning with L1-regularization. University of British Columbia, Canada

Schmidt M, Fung G, Rosaless R (2009) Optimization methods for L1-regularization. University of British Columbia, Canada

Sha F, Pereira F, Science I (2003) Shallow parsing with conditional random fields. In: Proceedings of ACL'03, vol 1, pp 134–141, ACL

Sutton C (2012) An introduction to conditional random fields. Found Trends® Mach Learn 4 (4):267–373

Yuan J, Zheng Y, Zhang C, Xie X, Sun G-Z (2010) An interactive-voting based map matching algorithm. In: 2010 Eleventh international conference on mobile data management pp 43–52

Road Network Conflation: An Iterative Hierarchical Approach

Andreas Hackeloeer, Klaas Klasing, Jukka Matthias Krisp and Liqiu Meng

Abstract Road Network Conflation is concerned with the unique identification of geographical entities across different road networks. These entities range from elemental structures such as crossings represented by nodes in the network to aggregated high-level entities such as topological edges or sequences of edges. Based on topological, geometrical and semantic information, the road networks to be conflated are investigated in order to identify similarities as well as differences. In this paper, we introduce a novel approach for conflating road networks of digital vector maps which iteratively employs multiple matching steps on different hierarchies of structures in order to progressively find, evaluate and refine possible solutions by recognizing and exploiting topological and geometrical relationships. The introduced algorithms are applied to real-world maps and validated against ground truth data retrieved from visual inspection. Validation shows that our approach leads to good results exhibiting high success rates in rural regions and provides a reasonable starting point for further refining in dense urban areas, where special heuristics are required in order to tackle difficult matching cases.

Keywords Road network conflation · Road network matching

A. Hackeloeer (✉) · L. Meng
Department of Cartography, Technische Universität München, Arcisstraße 21, 80333 Munich, Germany
e-mail: ahackeloeer@gmail.com

L. Meng
e-mail: liqiu.meng@bv.tum.de

K. Klasing
BMW Forschung und Technik GmbH, Hanauer Straße 46, 80992 Munich, Germany
e-mail: klaas.klasing@gmail.com

J.M. Krisp
Department of Geography, Universität Augsburg, 86135 Augsburg, Germany
e-mail: jukka.krisp@geo.uni-augsburg.de

G. Gartner and H. Huang (eds.), *Progress in Location-Based Services 2014*,
Lecture Notes in Geoinformation and Cartography,
DOI 10.1007/978-3-319-11879-6_10

1 Introduction

A road network is the structure given by the topology and geometry of roads, streets and transport links within a certain area. Historically, maps depicting road networks have a long tradition, dating back to a time as early as Ancient Egypt where maps such as the Turin Papyrus Map showed pedestrian routes along dry river beds (Harrel and Brown 1992). Nowadays advanced methods of georeferencing are employed in order to accurately assign the geographical objects used in a road map to geographical locations. Maps are increasingly authored and maintained in digital form, which may either be raster or vector based (Hackeloeer et al. 2014).

In recent times, digital vector maps have gained importance in the field of automotive navigation. These maps organize georeferenced information in several layers. The *topological layer* consists of a representation of the road network given by its induced graph, where crossings and intersections correspond to nodes in the graph, and the routes interconnecting these crossings correspond to edges. In contrast, the *geometrical layer* of a digital map contains a description of the geometrical shape of the objects stored in the map. While a road is usually represented as a one-dimensional entity, more detailed maps, e.g. those required for driver assistance systems, may store additional information such as the width of a road or the boundaries of different lanes. In addition, digital road maps may contain polygonal two-dimensional objects such as footprints of buildings or special areas like parking zones. All geographical entities found in a digital road map are georeferenced using a geodetic reference frame such as WGS-84 (Boucher and Altamimi 2001). Digital road maps usually enforce internal consistency, i.e. there is only one unique representation of a single geographical object. However, multiple maps of the same region may vary greatly, to the extent that a geographical object present in one map may be entirely missing in the other, or may be assigned to a different georeference. Also, it may (partly) correspond to multiple objects in the other map, e.g. if a road with lanes separated by a physical divider is modeled as a single road in one map and as two one-way roads in the other.

The problem of identifying geographical entities across different maps is called *conflation* (Saalfeld 1988). The outcome of a conflation process between two maps is a projection from one map to the other which defines how the geographical entities found in the two maps are related (the similarities), thereby also identifying the objects which are unrelated (the differences). Within the domain of conflation, *Road Network Conflation* is concerned with conflating road networks. With the advent of an increasingly heterogeneous landscape of location-based services which heavily rely on georeferencing, often across different maps and environments, Road Network Conflation is gaining attention as a path towards reliable attribute transfer, cross-map relating of georeferenced entities, and map fusion. Moreover, conflation offers a way to store georeferenced information independent of specific maps, which is of special importance in the automotive navigation, e.g. for maintaining learned georeferenced data in the course of a map update. This paper introduces the Road Network Conflation problem along with common approaches to solve the problem. Then, our novel approach called Iterative Hierarchical Conflation (IHC) is described, followed by a real-world map evaluation of the algorithm. Finally, we conclude by summarizing our results and discussing the ongoing challenges in the field.

2 Problem Definition of the Road Network Conflation Problem

Let $M_1 := (N_1, E_1)$ a graph representing a road network, where $E_1 \subseteq (N_1 \times N_1)$, and $M_2 := (N_2, E_2)$ another graph representing a road network of the same area, where $E_2 \subseteq (N_2 \times N_2)$. M_1 is also called the *Reference Network*, while M_2 is called the *Matching Network*. We call $N_i := \{N_i^1, \ldots, N_i^{n_i}\}, n_i \in \mathbb{N}$ the *nodes* of road network M_i and $E_i := \{E_i^1, \ldots, E_i^{m_i}\}, m_i \in \mathbb{N}$ the *links* of road network M_i. The nodes of both graphs are allowed to be bivalent, reflecting the fact that road networks of digital road maps often contain bivalent nodes, e.g. at places where a crossing existed in a prior version of the map or where a link-specific attribute such as a speed limit changes its value. We then call $S_i := (E_i^j, E_i^k, \ldots, E_i^x)$ a *link sequence* created from the concatenation of consecutive links of the edge relation E_i. A link sequence corresponds to a simple path from the starting node of E_i^j to the ending node of E_i^x.

We define a *node matching relation* $P \subseteq (N_1 \times N_2)$ as a set of node pairs, where for each pair the first node is taken from the nodes of M_1 and the second node is taken from the nodes of M_2. Thus, any $p = (p_1, p_2) \in P$ assigns a node $p_1 \in N_1$ to a node $p_2 \in N_2$. Note that in general, we neither require P to be functional, nor injective, i.e., one node of N_1 may be assigned to multiple nodes of N_2, and vice versa. This is motivated by the fact that many-to-many relationships between nodes may and are likely to exist when conflating road networks from different sources. A node matching relation represents a solution to the road network conflation problem on the elementary level of topological nodes.

In order to refine our solution towards the level of aggregated structures such as links and link sequences, we define a *link sequence matching relation* $L \subseteq (S_1 \times S_2)$ as a set of link sequence pairs, where for each pair the first link sequence consists entirely of consecutive links taken from E_1 and the second link sequence consists entirely of consecutive links taken from E_2. A link sequence matching relation represents a solution to the road network conflation problem on the level of one-dimensional structures. Again, many-to-many relationships between link sequences of the road networks to be conflated may exist. As indicated in the introduction, this may e.g. be the case if the modeling of a road with a physical divider differs between the road networks.

So far, we have only defined solutions which allow for describing total correspondences, i.e. situations where a link or a sequence of links of M_1 corresponds to another link or sequence of links of M_2 a whole. However, often partial

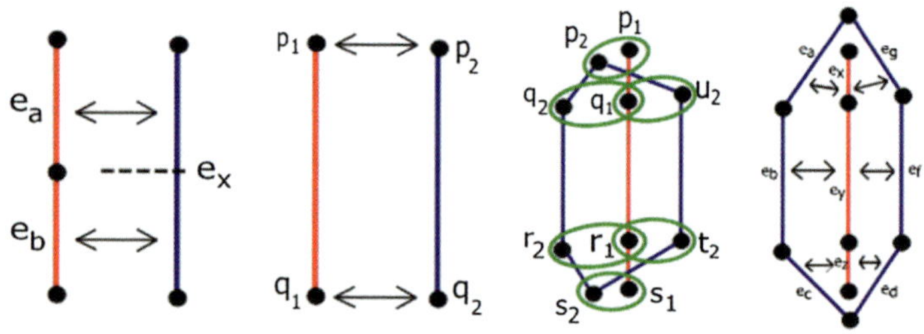

Fig. 1 From *left* to *right*: partial correspondence, 1:1 correspondence, one-to-many node correspondences, one-to-many link correspondences

correspondences are present. For example, it may occur that we have two links $e_a, e_b \in E_1$ and another link $e_x \in E_2$, and e_a corresponds to e_x only for the first couple of meters from the start of e_x, and the remainder of e_x corresponds to e_b in its entirety (see Fig. 1).

As a simple concept for dealing with partial correspondences, we suggest to employ *virtual nodes* along with *virtual links*. If a partial correspondence is identified, the involved links are split up into separate virtual links which are interconnected via a virtual node, and the virtual nodes and links are added to the respective road network. Node and link sequence matching assignments may then refer to these virtual entities in order to describe partial correspondences. Since a road network also includes a geometrical layer, some geometric modifications are necessary to update the geometry so as to match the altered topology. E.g., if the geometry of a link is given by a sequence of shape points, and the chosen split point is not identical to one of these shape points, then a new shape point must be created at an intermediate position along the straight line between the neighboring shape points. Formally, we need a set N_{i_v} of virtual nodes to be added to N_i, a set E_{i_v} of virtual links, where at least one of the two nodes in the link is virtual, and a projection $V_i : (E_i, N_{i_v}) \rightarrow (E_{i_v}, E_{i_v})$ which assigns a link involved in a partial correspondence to the two virtual links which are created from splitting the link at the position of the respective virtual node.

Taking these considerations into account, it is possible to express a solution to the Road Network Conflation Problem for two road networks on the level of both elementary topological nodes as well as composite line structures by the tuple $R = (P, L, N_{1_v}, E_{1_v}, V_1, N_{2_v}, E_{2_v}, V_2)$. The problem of Road Network Conflation can then be defined as finding the optimal R. Sometimes it is demanded that the conflation result exclusively describes one-to-one correspondences. In this case, we require both P and L to be injective as well as functional relations, and special strategies must be applied to deal with real-world ambiguities, such as replacing all nodes assigned to a node with a single merged virtual node which may e.g. be placed at their center of gravity.

3 Related Work

The following matching algorithms are approaches to the conflation of road networks: *Buffer Growing* (Walter 1997; Mantel and Lipeck 2004; Zhang and Meng 2007), *Multi-Stage Matching* (Xiong 2000; Volz 2006), and *Delimited Stroke Oriented Matching* (Zhang and Meng 2008; Zhang 2009).

3.1 Buffer Growing

Walter (1997) describes a geometric matching approach for line objects using the concept of Buffer Growing, which is also employed by Mantel and Lipeck (2004) for the matching of geometric datasets. Zhang and Meng (2007) suggest a road-matching approach based on Buffer Growing which also accounts for systematic geometric deviations by using unsymmetrical buffers.

Buffer Growing assumes a certain similarity regarding the location of the line objects to be matched, which may require preceding transformations. A line object originating from the Reference Network which we call the *reference link sequence* is encircled by a buffer, and then the Matching Network is spatially searched for all line objects which are fully covered by that very buffer, which we call *matching link sequences*. A result list holds assignments between reference link sequences and matching link sequences. In each iteration, the matching link sequence is added to the result list along with the corresponding reference link sequence as long as it cannot be derived from concatenating prior results, and then the reference link sequence along with the buffer boundary is extended by one link. The process stops when either certain pre-defined boundaries are reached, such as nodes having a valence (number of incident edges, where loops are counted twice) of 1 or a valence of at least 3, or if the reference link sequence exceeds a fixed number of links.

Buffer Growing already implies a limitation on the number of link sequence pairs to be evaluated by means of the size of the buffer. Still, the approach suffers from high combinatorial complexity, as no prior filtering derived from node assignments is performed.

3.2 Multi-stage Matching

Xiong (2000) proposes a three-stage approach for Road Network Conflation which consists of the stages node matching, segment matching, and edge matching. These stages are first processed bottom-up, i.e. from nodes via segments to edges, and then top-down, i.e. from edges via segments to nodes. In the bottom-up process, associations between nodes are established, which are then used to associate segments and edges. The edge mapping is aggregated from the segment mapping gained from associating the nodes. The top-down process propagates edge correspondences down to the level of elementary nodes in order to identify additional node associations. Volz (2006) describes an approach which relies on combined edge and node matching. After several preprocessing steps, seed nodes are identified in the Reference Network. Then, for each seed node, matching candidates are selected from the Matching Network. The process is repeated vice versa, i.e. with switched roles of the Matching and Reference Network. Once the seed nodes have been associated, the respective line objects, which correspond to link sequences, are compared and selected based on a number of topological and geometrical distance metrics. Finally, multiple iterations are performed in order to successively re-associate nodes by incorporating more tolerant criteria, which leads to the identification of new matching pairs.

Both approaches employ several heuristics and require fine-tuning of multiple parameters, so that tailored parameter settings are needed for a certain region. Also, they rely on a node matching algorithm which derives a similarity score between two nodes from the position of incident edges within discrete sectors. However, it has been shown that non-sector-oriented point matching algorithms lead to more accurate results (Hackeloeer et al. 2013).

3.3 Delimited Stroke Oriented Matching

Delimited Stroke Oriented Matching (DSO) is a Road Network Conflation algorithm introduced by Zhang (2009) which builds upon the Buffer Growing approach. After several preprocessing steps, a matching procedure is carried out which consists of five steps which are repeated at three different levels. The DSO algorithm operates on entities called *Delimited Strokes*, which are line objects corresponding to link sequences. In the matching process, first potential matching pairs are identified. Then, incorrect potential matching pairs are excluded by accounting for certain differences. The remaining matching pairs are subject to a further investigation, which removes some ambiguity by calculating a similarity score. A so-called *network-based selection* detects conjoint Delimited Strokes and arranges them into a single *network*, which is then used for *network-based matching*. Finally, the node pairs on twigs of the matched networks are used as seeds for matching-growing, which leads to the identification of new Delimited Stroke matching pairs. The growing continues as long as the new matched pairs exhibit sufficient geometrical and topological similarity. While the DSO algorithm is capable of dealing with numerous matching cases, the large number of heuristics involved leads to a very high overall complexity of the process in terms of both computation time as well as necessary parameter adjustments.

Table 1 Table of confusion for Road Network Conflation results

Algorithm decision/reality	True	False
Positive	Correctly identified matching pairs	Pairs incorrectly identified as being a match
Negative	Pairs correctly identified as being no match	Pairs incorrectly identified as being no match

4 Evaluation Methodology of Road Network Conflation

In order to find a solution, a conflation algorithm needs to perform a binary classification of pairs of nodes and links. Like any classifier, conflation algorithms can be assessed by means of predictive analytics. In detail, several properties derived from a table of confusion offer a starting point for the evaluation.

Lemma 1 A solution R is optimal if it maximizes the number of true positives and true negatives in both P and L, while minimizing the number of false positives and false negatives in both P and L (see Table 1).

A conflation algorithm is directed towards two rivaling goals: *correctness* and *completeness*. Correctness requires that the identified assignments reflect real-world correspondences, while completeness implies that existing real-world correspondences are actually identified as assignments. The extent to which correctness is achieved can be measured by the *precision* (sometimes called *positive predictive value*), while the degree of completeness may be expressed by the *recall* (also known as *sensitivity*). In this context, precision constitutes the percentage of correct algorithm decisions out of all algorithm decisions, and recall stands for the percentage of correct algorithm decisions out of all correspondences which are reflected by reality. Precision and recall are negatively correlated and thus cannot be optimized independently. It should be noted that the recall is not related to the actual *network coverage*, i.e. the percentage of elements which are part of the projection out of the number of all elements of a road network. If the road networks to be conflated offer few similarities, even an optimum solution with perfect precision and recall will exhibit little coverage for both networks. Since precision and recall for node and link solutions are directly correlated, it is sufficient to only evaluate link solutions in order to assess a conflation algorithm.

5 Iterative Hierarchical Conflation

Here, we introduce a novel approach to Road Network Conflation named Iterative Hierarchical Conflation, which combines concepts from both Multi-Stage Matching and Buffer Growing in order to find and iteratively refine matching results. From an

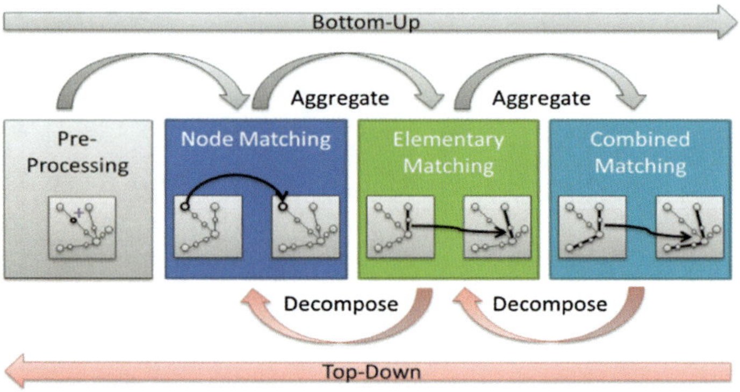

Fig. 2 The matching pipeline

abstract point of view, data are processed in the form of what we call the Matching Pipeline (see Fig. 2).

The Matching Pipeline basically consists of four stages: *Preprocessing*, *Node Matching*, *Elementary Matching*, and *Combined Matching*, where the latter may be repeated several times in order to find more assignments. The input of each stage is comprised of the output of all preceding stages, so that the matching result can successively be improved as more information regarding correspondences has been learned. Processing is divided into two phases: The *bottom-up* phase, where correspondences between elementary structures are aggregated, and the *top-down* phase, where correspondences between composite structures are decomposed.

5.1 Bottom-Up Phase

In the course of the bottom-up phase, correlations between elementary structures are identified, which are then combined in order to derive assignments between more complex structures.

5.1.1 Preprocessing

In order to normalize the road networks to be conflated, certain preprocessing must take place, depending on their deviance. For example, if there is a systematic geometric offset between both networks, it is possible to manually identify the offset and remove it so that the matching network is centered on the reference network. Also, it may be beneficial to harmonize the shape point resolution in each map to facilitate spatial queries. After normalization, index structures must be created which enable efficient spatial search for nodes as well as for shape points. This may e.g. be performed with a k-d-tree or a quadtree. If the road networks are

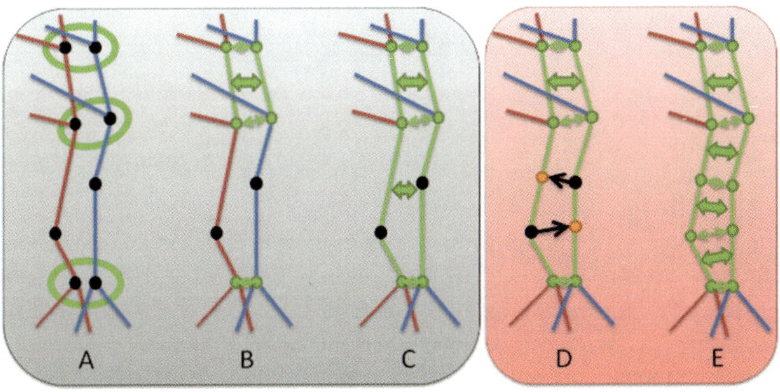

Fig. 3 Bottom-up phase (*left*) and top-down phase (*right*)

based on different data models, they must be converted so as to share the same data model in order to allow direct comparisons of their geometry and topology.

5.1.2 Node Matching

During bottom-up node matching, assignments between non-bivalent nodes of the road networks are identified (see Fig. 3A). While any point matching algorithm may be used for this task, we chose to employ the Exact Angular Index (EAI) approach introduced by Hackeloeer et al. (2013) as it provides several benefits in terms of precision compared to discrete sector-based point matching techniques such as the Spider Index. In detail, the EAI evaluates all possible projections from the edges of the reference node to the edges of the matching node and selects the projection which exhibits the lowest overall angle difference derived from aggregating the angle differences of all edge assignments, where redundant or missing edges are counted as worst-case angle differences of 180°.

In order to obtain a node matching solution P, a fixed-radius search in the set of non-bivalent nodes of the matching network is performed for each non-bivalent node p_1 of the reference network, resulting in candidate nodes $p_2^1 \ldots p_2^n$. Then, the EAI score is calculated for each pair (p_1, p_2^i). By always choosing the pair with the highest similarity score, the node matching solution comprises a functional relation, i.e. no node in the reference network is assigned to more than one node in the matching network. By repeating this process with switched roles of the networks (yielding an injective relation) and then intersecting both relation sets, a bijective node matching solution can be derived. This solution satisfies mutual optimality, i. e. for any pair (p_1, p_2) and a fixed radius r, p_1 is the best match for p_2 within a circle of radius r, and p_2 is also the best match for p_1 within a circle of radius r.

If all nodes which are part of the solution are removed from both networks, and then the process is repeated, additional node pairs may be identified. This can be done until no additional node pairs are found in an iteration, or until a pre-defined limit for the number of passes has been reached.

5.1.3 Elementary Matching

In the elementary matching stage, elementary link sequences, i.e. those consisting of only one link, are constructed starting from a given node matching solution. We establish two sets holding link sequences, one for each road network: $T_i := \{S_i^1, \ldots, s_i^{o_i}\}$. For each node of the reference network which is part of the solution, a separate link sequence is constructed out of each incident link, and all constructed link sequences are added to the corresponding link sequence set. The same process is repeated for the nodes of the matching network contained in the solution.

For performing the actual matching, the two link sequence sets are compared. For each link sequence $S_1^j \in T_1$, a corresponding link sequence $S_2^k \in T_2$ is identified if the start node of S_1^j is related to the start node of S_2^k in the node matching solution P, and also the end node of S_1^j is related to the end node of S_2^k. If a link sequence match between S_1^j and S_2^k has been found, the pair (S_1^j, S_2^k) is added to the link sequence matching relation $L \subseteq (S_1 \times S_2)$ (see Fig. 3B). By employing several index structures, this process can be turned into an $O(n+m)$ operation, where $n = |T_1|$ and $m = |T_2|$. In the next step, duplicates are removed from L (i.e. those pairs which are identical apart from having swapped start and end nodes). Finally, all link sequences of T_1 and T_2 are removed which have links in common with link sequences of L.

5.1.4 Combined Matching

The combined matching stage is concerned with the identification of correspondences between composite link sequences. Therefore, new composite link sequences are created by concatenating more elementary link sequences already present in both link sequence sets. A concatenation of two link sequences in a link sequence set T_i takes place if the end node of one link sequence is the start node of the other. In this case, a new link sequence is derived from the concatenation and added to the corresponding link sequence set. After concatenation has been performed for both link sequence sets, they are compared again in the same manner as during the elementary matching stage. As a result, new non-elementary link sequence pairs are added to the link sequence matching relation L (see Fig. 3C). This process may be repeated for a given number of passes, or until there is no further concatenation possible, which implies that no additional link sequences are created in an iteration.

L may contain ambiguity, i.e. one link sequence of the reference network may be assigned to multiple link sequences of the matching network, or vice versa. In order to enforce bijectivity and filter improbable matches, a score is assigned to each link sequence pair of L expressing the degree of similarity of their corresponding polylines, which is projected on the interval [0;1]. This score can be calculated using any polyline distance metric or a weighted combination of these metrics. In detail, a simple distance metric yielding good results is the length ratio between the polylines. Others include the sinuosity ratio, the Hausdorff distance, the Fréchet distance or the area of the enclosed polygon (Yuan and Tao 1999). Once the scores of all link sequence pairs have been determined, the pairs assigned to scores below a certain threshold score are removed, as it can be assumed that they do not reflect real-world correspondences. Then, L is made bijective in the same way as it has been done with the node matching result (see Node Matching).

If there exists topological inconsistency, which may e.g. be caused by incorrect point assignment in the node matching stage, multiple link sequence pairs of L share several, but not all links. In this case, it is not possible to establish a consistent matching and thus, link sequence pairs with common links are removed from L.

A special treatment is required in order to identify *dangling link sequence pairs*, i.e. pairs of link sequences which are only associated by their start nodes, but not by their end nodes. Such situations can occur if two roads are running in parallel for a certain distance, but beyond that one road ends while the other continues until it also reaches a dead end. These may reasonably be added to L if the end nodes are not associated to other link sequences. To create a proper link sequence pair out of a dangling link sequence pair, a corresponding virtual end node must be placed near the position of the end node of the shorter link sequence on the longer link sequence, which can then be associated with the non-virtual end node of the other sequence. Then, the same procedure as for regular link sequence pairs can be applied to take care of ambiguity and inconsistency. The dangling link sequence assignment must take place subsequent to all combined matching passes, since a dangling link sequence pair might be prolonged to a regular link sequence pair after concatenation has been done.

5.2 Top-Down Phase

During the top-down phase, correlations between aggregated structures are decomposed into more elementary associations.

5.2.1 Combined Matching

Over the course of the bottom-up phase, correspondences between link sequences have been identified. The implied knowledge that has been learned is the fact that the corresponding road segments refer to the same real-world entity, regardless of differences in topology and geometry. This knowledge can now be used to project nodes located on one link sequence onto a corresponding position on the other link sequence. Projection is done by multiplying the offset of a node from the start node of the link sequence by the length ratio between the two link sequences, then placing a virtual node at the resulting distance from the start node of the paired link sequence. This results in a splitting of the affected link into two new virtual links (see Fig. 3D). The underlying rationale is the assumption that the link sequence of the matching network as a whole represents a shrinked or stretched version of the entire corresponding link sequence of the reference network. In order to decide whether it is better to place a projected virtual node or rather associate with a nearby non-virtual node, a one-dimensional search can be performed within an interval around the projected position.

Sometimes multiple mappings are possible. Thus, we evaluate all possible projections from the nodes of the link sequence of the reference network to nodes of the associated link sequence of the matching network by aggregating and normalizing the EAI scores of the respective node pairs for every possible projection and then selecting the projection yielding the best overall score. A projection is deemed possible if there are no crossed assignments of nodes, as these cannot logically reflect real-world situations.

5.2.2 Elementary Matching and Node Matching

After the links of the link sequence pairs in $\boldsymbol{L}$ have been split according to node projections, it is now possible to establish an elementary mapping on the link level. Since for every node along a link sequence there is a matching node on the associated link sequence (either virtual or non-virtual), single-link pairings can be derived which represent total correspondences (see Fig. 3E). Finally, all nodes belonging to link sequence pairs that have not been correlated in the bottom-up phase are now added to the node matching result, including bivalent nodes.

6 Evaluation of the Iterative Hierarchical Conflation

6.1 Test Setup

Four sample regions were used: Two rural and two urban areas. As representatives for rural regions, we chose Moosach, Germany ([48.0335, 11.8729], [48.0292, 11.8801]) and Sullivan, NY, USA ([43, −75.79], [42.97, −75.735]). For the urban sample, we employed Munich Old Town, Germany ([48.138, 11.5738], [48.1349, 11.5804]) and Boston Financial District, MA, USA ([42.358, −71.062], [42.3533, −71.0553]). For the rural samples, we used a search radius of 40 m and a combined matching iteration limit of 2, and for the urban samples, 15 m and a limit of 5. For the road networks, we relied on map data from two different commercial map vendors who provide road maps for automotive navigation. In order to reduce the subjectivity inevitably involved with ground truth definitions, we employed a ground truth defined as the

Table 2 Summary of evaluation results for urban and rural regions

	Urban (Munich)	Urban (Boston)	Rural (Moosach)	Rural (Sullivan)
Found associations	121/134	121/140	78/82	53/55
False negatives	23	19	4	2
False positives	1	0	0	0
True negatives	10	28	10	4
Precision	99 %	100 %	100 %	100 %
Specificity	91 %	100 %	100 %	100 %
Recall	84 %	90 %	95 %	96 %
Nodes in network A	149	139	92	59
Nodes in network B	150	157	86	55
Network A coverage	81 %	87 %	85 %	90 %
Network B coverage	81 %	77 %	91 %	96 %

agreement of several people which we are disclosing here: https://www.dropbox.com/sh/6sox114wb6klx4h/AAAWhM1RnnLg1iIdjZCoU4ATa?dl=0

6.2 Results

Table 2 shows a summary of the evaluation results.

6.2.1 Results for Urban Samples

The conflation result for the Munich and Boston samples can be seen in Fig. 4. Most false negatives can be attributed to one-to-many or many-to-many correspondences

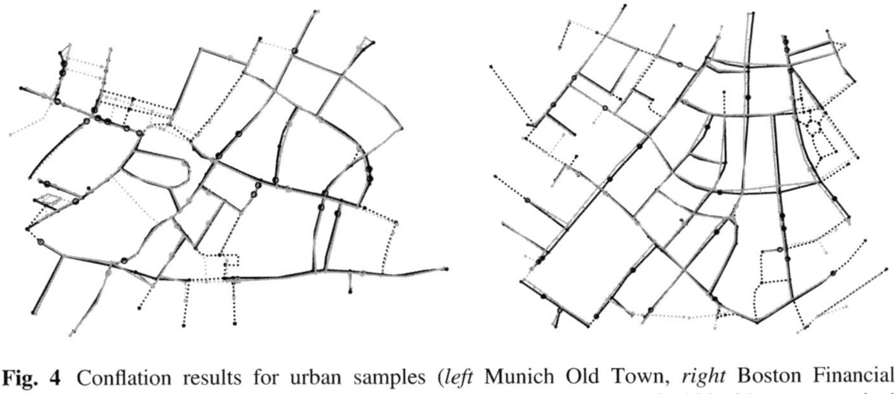

Fig. 4 Conflation results for urban samples (*left* Munich Old Town, *right* Boston Financial District). *Solid black/gray* matched links of corresponding network, *dashed black/gray* unmatched links. Node and link matchings are shown as *solid/thin dashed lines*. Nodes are shown as *black/gray dots*. Virtual nodes are encircled

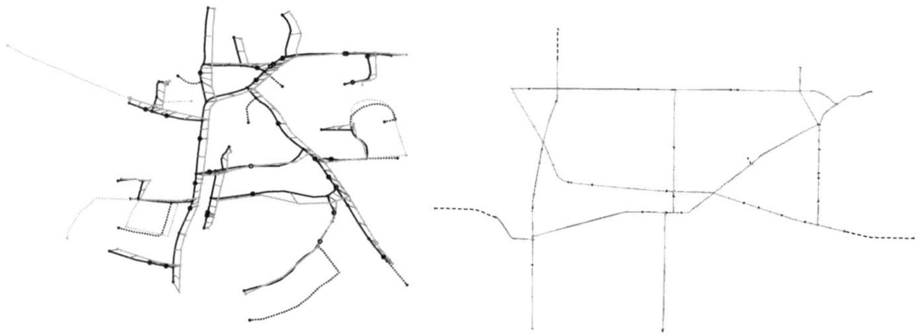

Fig. 5 Conflation results for rural samples (*left* Moosach, *right* Sullivan)

between nodes and links. The modest coverage indicates that there are several major differences between the networks, making the conflation process difficult and error-prone. It can be seen that the IHC algorithm is designed to deliver very reliable results by exploring topological relationships which are enforced for assignments, sometimes at the cost of missing some assignments which are solely related through geometry or cannot uniquely be deduced to a topological relationship.

6.2.2 Results for Rural Samples

Figure 5 shows a visualization of the conflation results for the rural samples.

For Moosach, the coverage suggests that the networks are fairly similar. The ground truth solution nearly matches the algorithmical solution, apart from a small area to the southwest. There, a topological inconsistency between the networks leads to an improper node assignment and thus to multiple link sequences sharing one link, which are removed during the bottom-up combined matching stage. The simple, but very large-scale Sullivan sample also exhibits a very good recall and perfect precision. The two networks do exhibit a shift, however it is nearly invisible due to the scale of the image.

7 Summary

In this paper, we have introduced the field of road network conflation. We gave a formal definition of the road network conflation problem as well as of the evaluation methodology for the assessment of road network conflation algorithms, which employs methods of the domain of predictive analytics. Furthermore, we described, discussed and classified common road network conflation approaches in the field. Subsequently, we presented our novel approach called IHC, which comprises a comprehensive multi-stage and bi-phase model which builds upon a combination of

Multi-Stage Matching and Buffer Growing in order to iteratively find correlations between geographical structures on different levels of aggregation. In the evaluation section, we assessed the correctness as well as the completeness of the IHC algorithm by performing a conflation of road networks provided by two commercial map vendors from four regions with different characteristics: two rural and two urban regions. We compared the conflation results with ground truth results derived from visual inspection and calculated precision and recall. Our results show that the IHC algorithm works very well in terms of both correctness and completeness in the rural sample regions and provides a very high correctness while maintaining considerable, but not perfect completeness in the urban sample regions. Thus, we conclude that further advancements of the IHC approach with special attention to the proper resolution of ambiguous correspondences are necessary to tackle hard matching cases such as the historic city center of Munich or Boston Financial District.

References

Boucher C, Altamimi Z (2001) ITRS, PZ-90 and WGS 84: current realizations and the related transformation parameters. J Geodesy 75:613–619

Hackeloeer A, Klasing K, Krisp JM, Meng L (2013) Comparison of Point Matching Techniques for Road Network Matching. International archives of the photogrammetry remote sensing and spatial information sciences, XL-2W1. pp 87–92

Hackeloeer A, Klasing K, Krisp JM, Meng L (2014) Georeferencing: a review of methods and applications. Ann GIS 20(1):61–69

Harrell JA, Brown VM (1992) The world's oldest surviving geological map—the 1150 BC Turin papyrus from Egypt. J Geol 100(1):3–18

Mantel D, Lipeck U (2004) Matching cartographic objects in spatial databases. ISPRS vol. 35, ISPRS Congress, Commission 4. Istanbul, Turkey

Saalfeld A (1988) Conflation: automated map compilation. Int J Geogr Inf Sys 2:217–228

Volz S (2006) An iterative approach for matching multiple representations of street data. Proceedings of ISPRS workshop on multiple representation and interoperability of spatial data. Hanover, Germany, pp. 101–110

Walter V (1997) Zuordnung von raumbezogenen Daten—am Beispiel der Datenmodelle ATKIS und GDF. DGK Reihe C, Nummer 480. München, Germany

Xiong D (2000) A three-stage computational approach to network matching. Transp Res Part C 8:71–89

Yuan S, Tao C (1999) Development of conflation components. Proceedings of the international symposium of geoinformatics and socioinformatics. Ann Abor, MI, USA

Zhang M (2009) Methods and implementations of road-network matching. Ph.D. Thesis, Technical University of Munich. Munich, Germany

Zhang M, Meng L (2007) An iterative road-matching approach for the integration of postal data. Comput Environ Urban Syst 31(5):597–615

Zhang M, Meng L (2008) Delimited stroke oriented algorithm working principle and implementation for the matching of road networks. J Geogr Inf Sci 14(1):44–53

Analysing the Usage of Spatial Prepositions in Short Messages

André Dittrich, Daniela Richter and Christian Lucas

Abstract Spatial prepositions such as *in*, *on* and *near* are important to describe where things are located in relation to other geographic features. Location-based services (LBS) usually disregard such spatial prepositions. Their automatic detection and interpretation is challenging, because prepositions are quite often used in non-spatial context (e.g., "in the afternoon"). This paper analyses spatial relations in short messages. Short messages typically have special characteristics (e.g., slang, abbreviations, etc.) and thus represent a special type of natural language. A sample corpus of short messages was used to extract descriptions based on spatial prepositions and to analyse their commonness of use. A frequency-based probability for each term to be spatial was calculated, which can serve as an indicator of a verbal spatial description and support the development of intelligent spatial language interpretation in automatic systems.

Keywords Spatial relations · Prepositions · Verbal spatial descriptions

1 Introduction

Our ability to navigate, give directions, and reason about spatial relationships is a key to describe spatial environments and to share information about it. The most common way to describe spatial scenes in natural language is through the use of prepositions like *in*, *on* or *near*. Such spatial prepositions reflect a scene depending

A. Dittrich (✉) · D. Richter · C. Lucas
Institute of Photogrammetry and Remote Sensing, Karlsruhe Institute of Technology, Karlsruhe, Germany
e-mail: andre.dittrich@kit.edu

D. Richter
e-mail: daniela.richter@kit.edu

C. Lucas
e-mail: christian.lucas@kit.edu

G. Gartner and H. Huang (eds.), *Progress in Location-Based Services 2014*,
Lecture Notes in Geoinformation and Cartography,
DOI 10.1007/978-3-319-11879-6_11

on the geometric arrangement and properties of the objects. According to Retz-Schmidt (1988), this concept involves at least two objects: a reference object (called relatum, ground, anchor or landmark) and a located object (called locatum, figure, trajector or theme) as well as the preposition itself. Throughout this paper, the terms relatum and locatum will be used.

The set of spatial prepositions in English counts to about 62 and is quite small compared to other word categories. However, computational detection of prepositions in a spatial meaning remains a difficult task. This problem intensifies when the natural language description is poor in syntax, is lacking conventional orthography, and is limited to a small set of characters. Such a situation is given for short messages on social media platforms (e.g., status updates on Facebook or micro blogs in Twitter). In the case of Twitter each message (tweet) is limited to a set of 140 characters. Millions of active daily users exchange information on all topics of everyday life (c.f. Java et al. 2007). That kind of short information is up-to-date and released directly from a potential point of interest. Due to these properties the tweets are, e.g., relevant in the domain of disaster management, which needs current and on site information about the affected area, passable roads, injured people etc. (c.f. Crooks et al. 2013). To take advantage of this information in near real-time an automatic evaluation of tweets, regarding the relevant information and especially spatial descriptions used, is required.

Apart from an application in the disaster management domain, textual represented spatial information from micro blogs is a new source for many location based information services. Such services are, for instance, the recommendation of social events coming from social media services or the real-time quality assessment revolving around restaurants, services, and other venues.

This paper will investigate the use of spatial prepositions in the context of short messages and identify some predominant cases on this basis. The next sections will introduce related work on spatial relations and prepositions (Sect. 2) and outline the research idea and the theoretical framework for the analysis (Sect. 3). In Sect. 4, a detailed analysis of the corpus and the classification of prepositions is presented. Sections 5 and 6 present and discuss the results of this analysis.

2 Related Work

A considerable amount of literature has been published on spatial relations. Levinson (2004), Tversky and Lee (1998) or Carlson and Van Deman (2004) started on a more general level analysing the role of space in language and cognition, i.e. the use of spatial relations expressed in language. Tversky and Lee (1998) pointed out that the meaning or sense of prepositions depends on functional aspects of the reference object. Considering the expression "on holiday", the noun *holiday* is not a locational reference, thus the potentially spatial preposition *on* is non-spatial. In contrast, the work of Herskovits (1980), Coventry and Garrod (2004) and Tenbrink (2005) concentrates on the role of spatial prepositions and their usage. Tenbrink (2005) showed that static spatial prepositions can be used in dynamic contexts. The underlying statement is consistent with Tversky and Lee (1998). As a result, the detection of a preposition in a spatial usage based only on keyword and substring matching methods is difficult. The problem was also analysed in more detail by Vasardani et al. (2012). Their work is focused on the spatial aspects of the more general preposition *at,* shown that most people use that preposition to locate themselves either in relation to buildings, or to confined outdoor areas at street level. In this sense prepositional phrases may be ambiguous in their classification. *At* can appear as a topological descriptor, projective term or distance-related term and further context is required for its interpretation.

Previous research has also focused on other spatial relations such as the prepositions *in* and *on*. Garrod et al. (1999) suggest to extend the interpretation of these prepositions to functional aspects such as containment and support. However, to the authors' best knowledge little attention has been paid to the individual frequency of spatial prepositions. Richter et al. (2013) demonstrated the use of spatial relations in combination with a classification of spatial granularity to disambiguate complex place descriptions. Using a small corpus of approximately 300 descriptions *in*, *on*, and *near* were the most frequent prepositions in their study (excluding *at*).

Despite some promising research in this area, spatial prepositions are often not taken into consideration in applications and are not processed as a part of the locative information. Kordjamshidi et al. (2010) implemented a concept, which deals with the extraction of spatial information from natural language sentences. This method is called spatial role labelling and facilitates mapping the terms onto formal spatial relations. Relatum, locatum and spatial relations are therein assigned to linguistic terms. Most current approaches for analysing spatial prepositions in natural language processing use part-of-speech tagging methods (e.g., (Zhang et al. 2009) or (Hall and Jones 2008)). Natural language information processing platforms, such as GATE (General Architecture for Text Engineering, (Cunningham 2002)), extract named entities and associated spatial relations in text, based on syntactical rules, which are trained or derived manually from a large-scale annotated corpus. The quality of the result depends on the type and style of the corpus, commonly consisting of newspaper articles, which have a regular sentence structure. In contrast, short messages have a sentence structure that is poor in syntax, lacks correct orthography and depends on the individual style of the user. The combination of space limitation and usual content leads to the use of acronyms, both standard and non-standard abbreviations and the trend to a telegraphic style as well as to fragmented or even ungrammatical sentences (c.f. Finin et al. (2010)). Especially in the field of natural language processing this poses new challenges. The informal and colloquial language carries completely different characteristics than the usual training corpora for part-of-speech tagging and related tools. Hence, to automatically extract spatial descriptions from short messages, this new corpus type needs to be analysed thoroughly.

3 Research Approach

Verbal spatial descriptions describe where things are located in space. In general, they consist of three different elements: The locatum, the relatum and their spatial relationship encoded by spatial prepositions, defining a region in which the locatum is located (c.f. Landau and Jackendoff 1993). A typical syntactical form of a verbal spatial description including a preposition is:

[[subject verb] preposition] NP.

Brackets indicate optional elements of the description. The noun phrase NP, representing the relatum, can be a simple noun ("station"), a compound ("Frankfurt station"), or a complex phrase aggregated from simpler noun phrases and relationships ("Frankfurt station, platform 1"). To illustrate this structure with an example, in the description "I'm near the station" the location of the subject "I" (representing the locatum) is described in relation to the NP "station" (the reference object) by the preposition *near*.

The scope of this paper is limited to English prepositions; it excludes utterances containing locative adverbs or verbs. Spatial descriptions using adverbs ("I'm downstairs", "the table is inside") specify the relatum only implicitly, and context knowledge is required for their correct interpretation. Verbs incorporating spatial relations (usually indicating a path) such as "to enter" or "to cross" can be paraphrased by a simpler verb plus a preposition (e.g., "to go into", "to go across"). As English can be classified as satellite-framed language, the paraphrased form is presumably more frequent. For the manual annotation we defined three rules for the decision process if a description is spatial or non-spatial (for our research purpose):

1. The triple of locatum—preposition—relatum can be clearly identified.
2. The locatum and relatum are real physical objects.
3. The preposition encodes the actual spatial relation between the locatum and relatum (i.e. statements such as "the thought in the back of my head" would not be included).

For convenience, we will use italics for spatial prepositions to express that the description fulfils the aforementioned constraints. Prepositional terms, which are part of such descriptions and the relations encoded by these prepositions, will be highlighted accordingly. Consequently, the cases where the constraints do not apply are labelled as non-spatial. We assume that a few spatial prepositional terms are most common or more frequent than others in verbal spatial descriptions in short messages. In particular, we aim to find an answer to the following questions:

- Which prepositional terms do people use in their verbal spatial descriptions?
- What are the most frequent prepositional terms, respectively which categories of spatial relations are predominant?

- Which prepositional terms have a high probability (frequency-based) to describe a spatial context in short messages?

Based on the findings, we will in further research develop machine-learning-based NLP (Natural Language Processing) algorithms to automatically identify and extract verbal spatial descriptions from short messages. The long-term aim is to develop functional models for GIS-based representations of these verbal spatial descriptions. With the focus on micro blogging the corpus data was collected from Twitter by automatically extracting a large number of messages that contain potentially spatial prepositional terms and a subsequent manual classification of the messages in spatial or non-spatial was conducted (cf. Sect. 4). The obtained results will in a first step reveal the general frequency of use of these defined terms, the absolute frequency of the *spatial* prepositional terms, as well as the frequency-based probability that a certain prepositional term represents a *spatial* relation. Moreover, the results are also presented for the categories that the prepositional terms can be assigned to. Following Tenbrink (2005) we use the categories:

a. **projective** (also spatial-dimensional) (e.g. *in front of, behind, right of, left of*)
b. **topological** (e.g. *in, on*)
c. **path-related** (e.g. *across, along*)
d. **distance-related** (e.g. *near, close to*)
e. **cardinal** —Prepositions that include the four cardinal directions (*north of, south of, west of, east of*) and the four intermediate directions (*northeast of, southeast of, northwest of, southwest of*). Finer graduations (e.g. *north–northeast*) are disregarded because of their rare use in natural language descriptions (c.f. Sect. 5).
f. **'at'** —*At* can be interpreted differently due to its ambiguity (cf. Vasardani et al. 2012). It can be classified as projective, topological or distance related. For example, without knowing the context, a person being "at the library" can be *inside* the library, *in front of*, or also just *near* the library building. Thus, we decided (based on our examples) to treat *at* as a separate group.
g. **other**— Prepositions that cannot be definitely assigned to one of the classes (a) to (f) based on their examples in the corpus.

The results will serve in future as a "gold-standard" for evaluating our automatic extraction and annotation procedure. All classification rules and critical cases were defined and discussed by a group of GI-researchers. As the presented approach and the findings are intended as a pre-investigation of this special corpus type, we refrained from inter-annotator analyses so far. In further steps, it is planned to integrate several additional annotators to evaluate and, if necessary, optimise our annotation rules.

4 Implementation

4.1 Corpus Collection

The presented study is based on a corpus of 3,005,848 short messages from the micro-blogging platform Twitter. The tweets were collected on three different days (1st April/May/June 2013) over 24 h periods starting on 10 pm UTC (i.e. between 4 pm and 6 pm local time), respectively. To avoid systematic effects, the corpus includes the beginning of the week (Monday), the time during the week (Wednesday) and the weekend (Saturday). The geographical area from which tweets were registered, was limited by a geographical bounding box in order to collect mainly messages from native English speaking users (northern limit 48° N, southern limit 24° N, western limit 127° W, eastern limit 112° W, i.e. eastern part of the USA). An additional constraint to the consideration for the final corpus was the automatic language identification by Twitter, which is provided with the tweets' metadata. Only messages with English as identified language were considered. However, automatic language detection is a complex task in computer linguistics and as such not completely error-free. The manual investigation of the samples yielded only 0.02 % messages from other languages. Hence, the number of these misinterpretations is statistically insignificant for the following frequency considerations. According to Java et al. (2007), the corpus can be assumed to have a random mixture of messages containing daily chatter, conversations, sharing of information or URL's and news. Considering these research results, we can assume that the corpus provides an adequate overview of the usage of natural language in short messages.

So far, the messages are collected and interpreted as single closed entities of information in this research, i.e. the courses of conversations were not tracked nor re-established through post-processing. Analysing the context of messages will probably be advantageous to identify implicitly provided reference objects, as it is the case when locative adverbs are used (Tenbrink 2005). These kind of utterances without an explicit relatum, however, are not in the scope of this paper.

4.2 Random Sample Sets

In the next step, the final text corpus described in Sect. 4.1 is used to extract example messages using at least one of the 62 prepositional terms in Table 3. This list was compiled from related research articles (cf. Landau and Jackendoff 1993; Holmes and Wolff 2013), English digital and online dictionaries, thesauri and

Table 1 List of prepositional terms with potential spatial meaning

Above	Back to	Far from	Near	Out (of)	Under
Across	Behind	Forth	Next (to)	Outside (of)	Underneath
Against	Below	From	North of	Outwith	Up
Ahead of	Beneath	In	Northeast of	Over	Upon
Along	Beside	In (the) back of	Northwest of	Right of	Via
Alongside	Between	In (the) front of	Of	South of	West of
Amid	Beyond	In the middle of	Off	Southeast of	Within
Amidst	By	In the midst of	On	Southwest of	
Aside	Close to	Inside (of)	On top of	Through	
At	Down	Into	Onto	To	
Atop	East of	Left of	Opposite (of)	Toward	

similar resources (Linguee,[1] Merriam-Webster[2] and WordNet[3]). The selection of terms is conducted according to the formal constraints outlined in Sect. 3. Some parts of the expressions may be optional (depicted in parentheses in Table 1) like in *next (to)* or *opposite (of)*. In these cases, both versions carry similar semantics to a large extent and can function as prepositions in English utterances. Others like *back to* or *ahead of* require their additional part to be full-fledged prepositions. So far, special linguistic forms to replace prepositions, such as "@" and "2" (instead of "at" and "to" respectively) are not included in the investigation but will be in future work.

Due to the enormous amount of messages, this filtering step is automated via a java program. The program scans all messages for instances of the following regex (regular expression) pattern.

"`(.*[\\s\\W])?`" + ***`prepositional term`*** + "`([\\s\\W].*)?`"

This pattern matches messages, which include the string representation of the respective prepositional term preceded and followed by a space, any non-word character (e.g., #:* ~//!? etc.) or no character at all. Depending on the prepositional term, this approach may lead to the matching of homonymous terms (cf. Sect. 4.3). The program does not incorporate any spell checking metrics, i.e. no misspelled version of the searched terms will be detected. In contrast, in particular cases where the misspelling of a word leads to the accordance with one of the search terms, this message would be included (cf. Sect. 4.3). The identified example messages are compiled to form collections based on the matched prepositional term for manual classification. If a message contains several prepositional terms, it will occur in all the respective collections ("...Tornado Warning *in* Maine heading *from* Kingfield

[1] Linguee. From http://www.linguee.de. Accessed 12. February 2,014.

[2] Merriam-Webster. From http://www.merriam-webster.com. Accessed 12. February 2,014.

[3] WordNet (Software): http://wordnet.princeton.edu/wordnet/download/.

to just *north of* Bingham…"). Thus, it can be classified separately for each prepositional term.

On the whole, the program extracted a subset of nearly 2 million examples for the classification step (cf. Table 3). As this amount cannot be classified manually within an acceptable time frame, we decided to select random samples. These sample sets contain 100 messages (if applicable) for each of the search terms. This constraint resulted in close to 5,000 messages for manual classification (cf. Table 3). Those samples form the basis to estimate the absolute number of instances with *spatial* meaning for each term in the corpus.

4.2.1 Statistical Aspects

The estimation for the real proportion p of the *spatial* instances of a term in the corpus is given by

$$\hat{p} = \frac{n_{spatial}}{n} \tag{1}$$

n size of the random sample.
$n_{spatial}$ number of identified *spatial* instances in the random sample.

Thus, we estimate the absolute number of *spatial* instances of a search term in the corpus as

$$\hat{N}_{spatial} = N \cdot \hat{p} \tag{2}$$

N number of all instances of the term in the corpus.

The standard error of the estimate $\hat{p}$ relating to a confidence coefficient z can be modelled as

$$s_{\hat{p}} = \pm z \sqrt{\frac{N-n}{nN} \hat{p}(1-\hat{p})} \tag{3}$$

In case of the absolute estimate $\hat{N}_{spatial}$, the standard error is given by

$$s_{\hat{N}_{spatial}} = N \cdot s_{\hat{p}} \tag{4}$$

To provide a confidence level of 90 % assuming a normal distribution, z is set to 1.6449.

4.3 Manual Classification

The extracted subset of messages described in the previous section was classified manually in to *spatial* and *non-spatial*. Here, *spatial* is used for descriptions that contain *spatial* prepositions as defined in Sect. 3. In contrast, descriptions were classified as *non-spatial* if they did not comply with our definition. Three main cases are distinguished within this group:

1. **temporal**—Descriptions with a temporal meaning, e.g., "at 6 pm".[4] By default prepositions that may indicate both temporal and spatial relationships (e.g., "within 10 s", "within walking distance of our new apartment") are also classified as *non-spatial*. Only if they contained a reference object, as in the latter example, they would be classified as *spatial*").
2. **symbolic/figurative**—Descriptions, such as "I'm beside myself", "I will keep her inside my heart", "doubts along the way" or "the wind under my wings" were classified as *non-spatial*, because the alleged spatial relation does not actually exist.
3. **adverbial use**—Descriptions incorporating adverbs, e.g., "I went downstairs" are excluded from the class *spatial*, because the reference object is not explicitly mentioned in the defined structure (locatum-preposition-relatum).

During the classification of *spatial* and *non-spatial* relations, some further issues were discovered that require special consideration. There are certain terms, which may be used as prepositions as well as adjectives, nouns, proper nouns, etc. The examples "you are right of course" and "Right of way" contain the search string *right of* that may indicate a *spatial* relation, but is used as a noun in these cases. Furthermore microblogs are typically short and often use an abbreviated, colloquial and informal language (e.g., "northern KY" instead of the official place name "Northern Kentucky"). Also, misspellings (e.g., written "along" but meant "a long") may occur more frequently. By manual classification, such issues can be identified and resolved. Ambiguities in language are a further difficulty. As mentioned in Sect. 4.2, the automatic extraction can yield homonymous terms. The preposition *left of*, for instance, can also mean *what remains of something*. In the sentence "He took what was left of my soda and threw it at the window." *what* may refer to the remaining soda in a glass, but also to something located left of it. In this particular case we decided to classify *left of* as *non-spatial*. The phrase "in the movies" may be interpreted spatially (physically being in the cinema) in contrast to, e.g., actors playing in the movie. Any such issues were considered individually and classified according to the more common interpretation.

All *spatial* prepositions were furthermore categorized as described in Sect. 3 into projective, topological, path-related, distance-related, cardinal, and the separate

---

[4] All given text samples in this section are taken from the corpus in their original spelling.

group 'at'. Terms that were not used as *spatial* prepositions in our corpus were classified as 'undefined'. This classification structure is used for a clear representation of our examples.

5 Results

In this section, we will present the results of our automatic extraction and the manual classification of the sample sets. In Table 3, we provide a comprehensive overview of all search terms and their individual statistics as well as some overall statistics. The table is arranged in descending order of $\hat{N}_{spatial}$, i.e. the estimated number of *spatial* instances of the respective term in the corpus. In sum, our automatic extraction yielded 1,937,255 examples containing at least one of the search terms. Out of the total of 4,986 messages randomly chosen for manual classification, 28 % (i.e. 1,408) were classified as *spatial*. Thus, an undifferentiated approach over all terms would lead to the assumption of an approximate of 547,063 ± 20,289 verbal *spatial* descriptions in the corpus. The summation of the individual statistics, however, yields a more realistic estimate of 445,265 ± 106,031, leading to an approximate percentage of 23 %.

Besides, the table reveals that the first 20 terms, i.e. about one third of all terms, already form 98.8 % of all identified *spatial* examples in our corpus. Furthermore, the first 6 terms (*at*, *in*, *to*, *on*, *from* and *out (of)*, i.e. 10 % of all terms) yield 90 %, and the term *at* alone is already responsible for 34.6 % of all *spatial* prepositions in our corpus. With respect to the total corpus (3,005,848 messages), only the terms *at*, *in*, *to* and *on* occur as part of verbal *spatial* descriptions in more than 1 % of all messages. In contrast, 35 terms occur in less than 0.01 %.

Another interesting aspect is presented in the last column, depicting the probability of a term being used in a spatial context, based on the frequency in the sample set. Apart from the cardinal directions, only the terms *in (the) front of*, *across*, *at*, *beside* and *underneath* exceed a probability of 75 % (including their 90 % uncertainty range). As expected, these statistics qualitatively follow Zipf's law. However, important for our further research are the actually retrieved absolute and relative frequencies for the single prepositions.

Table 2 follows the same order as Table 3, but the terms are aggregated according to the category they are assigned to. With its estimated magnitude ($\hat{N}_{spatial}$), the separate category *'at'* contains the second-most *spatial* examples (154,164), outnumbered only by the path-related terms (205,774). In contrast, the category of cardinal directions comprises only 123 examples conveying *spatial* meaning.

Figure 1 on the other hand shows the probability of a term from a certain category to be *spatial*. In this representation, the category of the cardinal directions has the most *spatial* instances (97 %). Except for the category 'at', all other

Table 2 Accumulated statistics fort the categories of prepositional terms

Category	Terms	N	n	$n_{spatial}$	$\hat{N}_{spatial} \pm s_{\hat{N}_{spatial}}$	$\hat{p} \pm s_{\hat{p}}[\%]$
Path-related	14	888,149	1,338	310	205,774 ± 16,838	23 ± 2
'At'	1	205,552	100	75	154,164 ± 14,637	75 ± 7
Topological	9	472,884	650	153	111,310 ± 12,934	24 ± 3
Projective	15	44,231	1,370	576	18,596 ± 955	42 ± 2
Distance-related	5	68,868	500	130	17,906 ± 2,214	26 ± 3
Other	4	42,630	400	41	4,370 ± 1,058	10 ± 2
Cardinal	8	127	127	123	123 ± 0	97 ± 0

categories stay below 50 % of *spatial* instances. This means on the other hand that for those groups more instances are *non-spatial* rather than *spatial.*

In case of the individual terms, only 13 out of the 62 search terms yielded a higher frequency of *spatial* instances than *non-spatial* ones.

6 Discussion and Outlook

Our hypothesis from the outset of this paper was that certain *spatial* prepositional terms are used predominantly in short messages. The results in Sect. 5 clearly support this assumption and provide quantitative values for the intended automatic classification in the future.

A small set of terms, namely *at, in, to, on, from* and *out (of)*, account for the majority (90 %) of all verbal *spatial* descriptions in the corpus, whereas several other prepositional terms hardly occur at all to describe a *spatial* meaning in the messages. The predominant terms all provide a rather imprecise description of the actual *spatial* relation compared to more complex terms, such as *underneath, on top of, in the middle of* or the cardinal directions. We assume that this mapping of spatial relations onto more general and simple terms is due to the special characteristics of short messages, i.e. their generally simple language structure (or even ungrammatical sentences) and their character limitation. The answer concerning the frequency-based probability of a specific prepositional term to be *spatial* when it occurs in short messages can also be deducted from Table 3. As expected, the four cardinal directions and *southwest of* (the other intermediate directions did not occur in the corpus) each have a probability close to 100 %. Other terms with a value reaching at least 75 % (including the respective confidence range) are *in (the) front of, across, at, beside* and *underneath.*

Implications for our long-term goal, the development of machine-learning-based NLP-algorithms to automatically identify, extract, interpret and link verbal *spatial* descriptions in short messages, can be inferred from these findings. The estimated percentage of only 23 % *spatial* instances over all terms and the great variability

Table 3 Overview of all search terms and their individual statistics

Term	Category	N	n	$n_{spatial}$	$\hat{N}_{spatial} \pm s_{\hat{N}_{spatial}}$	$\hat{p} \pm s_{\hat{p}}(\%)$
At	At	205,552	100	75	**154,164 ± 14,637**	75 ± 7
In	Topological	268,168	100	33	**88,495 ± 20,738**	33 ± 8
To	Path-related	553,523	100	15	**83,028 ± 32,508**	15 ± 6
On	Topological	193,498	100	28	**54,179 ± 14,287**	28 ± 7
From	Path-related	54,369	100	27	**14,680 ± 3,967**	27 ± 7
Out (of)	Path-related	93,498	100	7	**6,545 ± 3,922**	7 ± 4
Back to	Path-related	9,793	100	53	**5,190 ± 800**	53 ± 8
Down	Path-related	27,754	100	17	**4,718 ± 1,712**	17 ± 6
Into	Path-related	15,260	100	29	**4,425 ± 1,135**	29 ± 7
Off	Other	35,572	100	11	**3,913 ± 1,828**	11 ± 5
By	Distance-related	44,990	100	8	**3,599 ± 2,005**	8 ± 4
Next (to)	Distance-related	20,038	100	16	**3,206 ± 1,205**	16 ± 6
Over	Projective	31,747	100	8	**2,540 ± 1,414**	8 ± 4
Up	Path-related	117,227	100	2	**2,345 ± 2,698**	2 ± 2
Through	Path-related	10,516	100	19	**1,998 ± 675**	19 ± 6
In (the) front of	Projective	1,927	100	89	**1,715 ± 97**	89 ± 5
Under	Projective	3,376	100	43	**1,452 ± 271**	43 ± 8
Behind	Projective	3,483	100	41	**1,428 ± 278**	41 ± 8
Across	Path-related	1,657	100	77	**1,276 ± 111**	77 ± 7
Near	Distance-related	2,018	100	49	**989 ± 162**	49 ± 8
Inside (of)	Topological	2,852	100	32	**913 ± 215**	32 ± 8
Outside (of)	Topological	6,384	100	12	**766 ± 339**	12 ± 5
Between	Other	3,620	100	14	**507 ± 204**	14 ± 6

(continued)

Table 3 (continued)

Term	Category	N	**n**	$n_{spatial}$	$\hat{N}_{spatial} \pm s_{\hat{N}_{spatial}}$	$\hat{p} \pm s_{\hat{p}}(\%)$
Beside	Projective	522	100	74	**386 ± 34**	74 ± 6
Close to	Distance-related	1,455	100	25	**364 ± 100**	25 ± 7
Against	Other	2,969	100	12	**356 ± 156**	12 ± 5
In the middle of	Topological	1,058	100	33	**349 ± 78**	33 ± 7
Above	Projective	913	100	23	**210 ± 60**	23 ± 7
On top of	Projective	507	100	32	**162 ± 35**	32 ± 7
Toward	Path-related	1,208	100	13	**157 ± 64**	13 ± 5
Along	Path-related	2,079	100	7	**146 ± 85**	7 ± 4
In (the) back of	Projective	203	100	69	**140 ± 11**	69 ± 5
Onto	Path-related	578	100	22	**127 ± 36**	22 ± 6
Far from	Distance-rel ated	367	100	32	**117 ± 24**	32 ± 7
Below	Projective	366	100	32	**117 ± 24**	32 ± 7
Underneath	Projective	163	100	71	**116 ± 8**	71 ± 5
Within	Topological	874	100	12	**105 ± 44**	12 ± 5
Upon	Path-related	649	100	10	**65 ± 29**	10 ± 5
Beneath	Projective	99	99	63	**63 ± 0**	64 ± 0
North of	Cardinal	40	40	38	**38 ± 0**	95 ± 0
South of	Cardinal	37	37	36	**36 ± 0**	97 ± 0
Ahead of	Projective	484	100	6	**29 ± 17**	6 ± 3
East of	Cardinal	26	26	25	**25 ± 0**	96 ± 0
West of	Cardinal	22	22	22	**22 ± 0**	100 ± 0
Opposite (of)	Other	469	100	4	**19 ± 13**	4 ± 3
Atop	Projective	30	30	12	**12 ± 0**	40 ± 0

(continued)

Table 3 (continued)

Term	Category	N	**n**	$n_{spatial}$	$\hat{N}_{spatial} \pm s_{\hat{N}_{spatial}}$	$\hat{p} \pm s_{\hat{p}}(\%)$
Alongside	Path-related	38	38	12	**12 ± 0**	32 ± 0
Right of	Projective	41	41	12	**12 ± 0**	29 ± 0
Left of	Projective	370	100	1	**4 ± 5**	1 ± 1
Southwest of	Cardinal	2	2	2	**2 ± 0**	100 ± 0
Amidst	Topological	6	6	1	**1 ± 0**	17 ± 0
Amid	Topological	9	9	1	**1 ± 0**	11 ± 0
In the midst of	Topological	35	35	1	**1 ± 0**	3 ± 0
Aside	Undefined	289	100	0	**0 ± 0**	0 ± 0
Beyond	Undefined	1,550	100	0	**0 ± 0**	0 ± 0
Forth	Undefined	352	100	0	**0 ± 0**	0 ± 0
Of	Undefined	209,179	100	0	**0 ± 0**	0 ± 0
Out with	Undefined	1	1	0	**0 ± 0**	0 ± 0
Via	Undefined	3,443	100	0	**0 ± 0**	0 ± 0
Northwest of	Cardinal	0	0	0	**0 ± 0**	0 ± 0
Northeast of	Cardinal	0	0	0	**0 ± 0**	0 ± 0
Southeast of	Cardinal	0	0	0	**0 ± 0**	0 ± 0
Σ		1,937,255	4,986	1,408	**445,265 ± 106,031**	

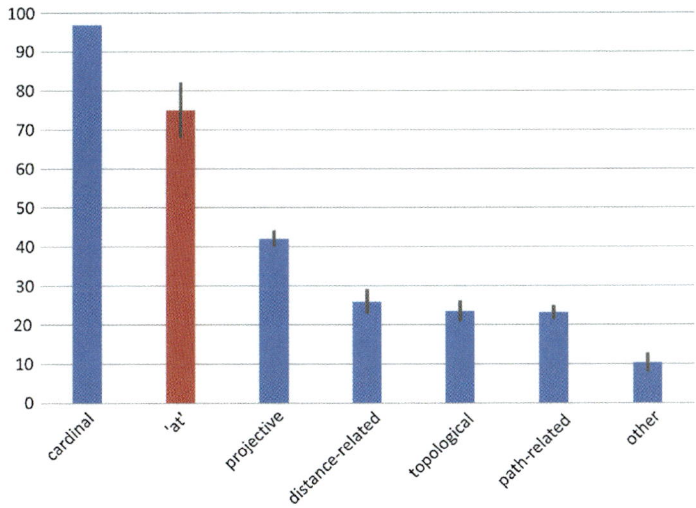

Fig. 1 Estimated percentage of spatial instances of a category in the corpus ($\hat{p}$). *Black* bars indicate the estimation error $s_{\hat{p}}$

between the individual terms concerning their probability of being *spatial* indicate that the benefits of an investigation of each term would not justify the expenses. In contrast, the predominant use of specific prepositional terms in verbal *spatial* descriptions suggests a more focused approach. Therefore, in future research, we will concentrate on the most common terms identified in this paper. Thus, a large percentage of verbal *spatial* descriptions given in short messages can already be covered. During such further research, the manually classified examples of these prioritized terms will serve as training set with positive and negative cases for a machine learning approach. The individual probabilities of the terms to be *spatial* can also be incorporated in an automatic system—initially as one indicator for a verbal *spatial* description and also as part of an estimation to express how likely the classification of an utterance as *spatial* really is. Moreover, we expect that terms with a high frequency-based probability to be *spatial* also have a better chance to be automatically identified correctly through NLP-approaches. Although the results of this paper are highly dependent on the manual classification, they present a first step to evaluate the automatic identification and extraction of *spatial* descriptions. The results, e.g., can help to build a hierarchical structure where rarely used terms are aggregated under a collective term based on their usage. Thus enhancing the coverage of identified *spatial* descriptions. Certain words that also encode spatial relations, i.e. locative adverbs and path-indicating verbs, were not considered in this paper. They usually lack an explicit reference object (locative adverbs) or are rarely used in English utterances (path-indicating verbs). However, in further research their investigation could lead to a more comprehensive identification of verbal *spatial* descriptions. In case of locative adverbs, the determination of the implicit relatum might be achieved through dialogue analysis or the consideration of context knowledge.

Eventually, a more detailed classification into the different non-spatial groups (temporal, figurative, adverbial use) might be advantageous for the automatic exclusion of descriptions that are *non-spatial*. Especially, a large amount of the utterances with figurative spatial sense and special cases could be covered by the means of a dictionary approach incorporating a list of patterns of English sayings such as "I'll keep you in my heart".

This research was concerned only with short messages; therefore, future work should evaluate the applicability for other sources as well.

7 Conclusion

This research demonstrates that there are a few prepositional terms that are clearly used predominately in verbal *spatial* descriptions in short messages. Several others apparently hardly occur at all in a *spatial* sense in this kind of utterances. We motivated to focus on the predominantly used terms for a machine-learning approach and already gathered a large training set of *spatial* and *non-spatial* examples through the manual classification. Additionally, we provided a frequency-based probability for each term to be *spatial* and suggest its usage as an indicator of a verbal *spatial* description in automatic systems.

Finally, we pointed out some future research directions that will support the development of NLP-systems to automatically identify, extract, interpret and link verbal *spatial* descriptions in short messages.

References

Carlson LA, Van Deman SR (2004) The space in spatial language. J Mem Lang 51(3):418–436

Coventry KR, Garrod SC (2004) Saying, seeing and acting: the psychological semantics of spatial prepositions. Psychology Press, U K

Crooks A, Croitoru A, Stefanidis A, Radzikowski J (2013) #Earthquake: Twitter as a distributed sensor system. Trans GIS 17(1):124–147

Cunningham H (2002) GATE, a general architecture for text engineering. Comput Humanit 36 (2):223–254

Finin T, Murnane W, Karandikar A, Keller N, Martineau J, Dredze M (2010) Annotating named entities in Twitter data with crowdsourcing. Proceedings of the NAACL HLT 2010. California Association for Computational Linguistics, Los Angeles. pp 80–88

Garrod S, Ferrier G, Campbell S (1999) In and on: investigating the functional geometry of spatial prepositions. Cognition 72:167–189

Hall M, Jones C (2008) Quantifying spatial prepositions: an experimental study. ACM GIS Irvine, CA, USA

Herskovits A (1980) On the spatial uses of Prepositions. 18th annual meeting of the association for computational linguistics Philadelphia, Pennsylvania, USA

Holmes KJ, Wolff P (2013) When is language a window into the mind? Looking beyond words to infer conceptual categories. 35th annual conference of the cognitive science society, Austin, TX

Java A, Song X, Finin T, Tseng B (2007) Why we twitter: understanding microblogging usage and communities. Proceedings of the 9th WebKDD San Jose, California ACM. pp 56–65

Kordjamshidi P, van Otterlo M, Moens M-F, Kordjamshidi P (2010) From language towards formal spatial calculi. Proceedings of the workshop on computational models of spatial language interpretation at spatial cognition, Oregon, USA

Landau B, Jackendoff R (1993) "What" and "where" in spatial language and spatial cognition. Behav Brain Sci 16(2):217–238

Levinson SC (2004) Space in language and cognition. Cambridge University Press, Cambridge

Retz-Schmidt G (1988) Various views on spatial prepositions. AI Magazine 9:2

Richter D, Winter S, Richter K-F, Stirling L (2013) Granularity of locations referred to by place descriptions. Comput Environ Urban Syst 41:88–99

Tenbrink T (2005) Semantics and application of spatial dimensional terms in English and German. Collaborative Research Center SFB/TR 8

Tversky B, Lee PU (1998) How space structures language. An interdisciplinary approach to representing and processing spatial knowledge. Spatial cognition. Springer, Berlin

Vasardani M, Winter S, Richter K-F, Stirling L, Richter D (2012) Spatial interpretations of preposition "at". Proceedings of the 1st ACM SIGSPATIAL, California ACM. pp 46–53

Zhang C, Zhang X, Jiang W, Shen Q, Zhang S (2009) Rule-based extraction of spatial relations in natural language text. Proceedings of the International Conference on Computational Intelligence and Software Engineering (CiSE), Wuhan, China

Using Location-Based Social Media for Ranking Individual Familiarity with Places: A Case Study with Foursquare Check-in Data

Wangshu Wang

Abstract The growing popularity of location aware social media provides a unique opportunity to study individual knowledge of the environment, e.g., individual familiarity with places. With Foursquare being one of the most popular location-based social media, in this paper, we focus on ranking individual familiarity with places using Foursquare check-in data. Our method firstly identifies individually meaningful places. Then, the identified meaningful places are ranked according to individual's familiarity with them, i.e., the weighting that we assigned to each place based on the information indicated by the tagging activities. Results of the evaluation demonstrate the possibility of ranking individual familiarity with places using location-based social media.

Keywords Place identification · Location-based social media · Spatial cognition

1 Introduction

As human understanding of the environment begins with places (Shemyakin 1962), deriving meaningful places (i.e., places that are associated with certain activities and meanings) and individual familiarity (i.e., how familiar are these places to an individual) with them from different sources of geographic data is always an interesting topic to researchers. The blooming of location-based social media in recent years brings us to a new information era, which generates tremendous geographic data and offers the possibility to study individual familiarity with places.

Existing researches on location-based social media focus on the aspects of mining place semantics and social knowledge (Rattenbury et al. 2009; Hollenstein and Purves 2010), identifying city landmarks and neighborhood boundaries (Chen et al. 2009; Keßler et al. 2009), estimating geographical location of a photo

W. Wang (✉)
Research Group Cartography, Vienna University of Technology, Vienna, Austria
e-mail: e1129622@student.tuwien.ac.at

G. Gartner and H. Huang (eds.), *Progress in Location-Based Services 2014*,
Lecture Notes in Geoinformation and Cartography,
DOI 10.1007/978-3-319-11879-6_12

(Hays and Efros 2008), studying people's travel behaviors (Zheng et al. 2011), understanding human and crowd mobility (Cheng et al. 2011; Naaman et al. 2012), interpreting human activities and city dynamics (Noulas et al. 2011; Cranshaw et al. 2012) and so on. However, to the best of our knowledge, none of them has paid attention on individual. While several researchers leverage GPS (Global Positioning System) trajectories or GSM (Global System for Mobile Communications) data to derive individually meaningful places (Ashbrook and Starner 2002; Zhou et al. 2007; Nurmi 2009), they have not gone further to individual familiarity with these places. Therefore, a study of the possibility and method to model individual familiarity with places is in need.

Our study aims at ranking individual familiarity with places using Foursquare check-in data, which is a starting point to model individual knowledge of places. This helps to provide location-based services adapted to individual priori spatial knowledge and assists us to better understand people's spatial knowledge and environmental perception.

To achieve the general goal, we focus on two aspects. The first one is the identification of individually meaningful places. The second one is ranking individual familiarity with these places.

This paper firstly reviews the researches on place identification using location-based social media and discovering individually meaningful places using different geographic data sources. Following a discussion and comparison of the clustering methods to identify individually meaningful places in Sect. 3, we propose and evaluate an approach to rank individual familiarity with places using Foursquare check-in data in Sect. 4. In the last section, we state our findings and draw the conclusion.

2 Related Work

2.1 Place Identification Using Social Media

With the outbreak of location-based social media, more and more people start to share their locations on the platforms, like Foursquare,[1] Twitter,[2] Flickr,[3] and Facebook.[4] On these platforms, geographic information is shared either explicitly as check-in or implicitly as location coordinates along with user post. The huge amount of geographic data is easily accessible, which provides a new data source to study people's spatial knowledge.

A lot of studies have been brought about on the different aspects of extracting and interpreting geographic information using location-based social media.

[1] https://foursquare.com/.

[2] https://twitter.com/.

[3] http://www.flickr.com/.

[4] https://www.facebook.com/.

Kennedy et al. (2007) and Rattenbury et al. (2009) explored place and event semantics of georeferenced tags. Zheng et al. (2011) looked into the geotagged photos on Flickr to analyze people's travel behavior of a tourist destination. Also based on geo-tagged photos, Hays and Efros (2008) developed a system to estimate the geographic location of a photo.

Based on the enormous georeferenced data, identification of city landmarks and neighborhood boundaries has also been investigated a lot. Chen et al. (2009) leveraged geo-tagged photos to generate landmarks of an area. Keßler et al. (2009) proposed a bottom-up approach to build gazetteer based on geotagged photos. They firstly used a crawling algorithm to retrieve geotagged photos with certain annotation. Then, a Delaunay triangulation was performed within point clouds to find clusters. Their result demonstrates the possibility to derive meaningful places from geotagged photos. Similarly, Flickr (2008) itself also practiced on deriving neighborhood boundaries from geotagged photos. Instead of Delaunay triangulation, the algorithm they used was Alpha Shape (Edelsbrunner et al. 1983).

As one of the most popular location-based social media, Foursquare has been utilized to study human behavior in different cities (Noulas et al. 2011; Cranshaw et al. 2012), in which the user history and some of the parameters inside a check-in were leveraged to understand human spatial knowledge non-intrusively.

However, their focuses are on large-scale place identification and conventionally meaningful places, while nearly no research has been found on identifying individually meaningful places using social media.

2.2 Discovering Individually Meaningful Places

Early studies on individual place identification (Marmasse and Schmandt 2000; Ashbrook and Starner 2002) only take physical location into consideration. The term "place" there is more likely to refer to location without specific meaning.

To bridge physical locations that computers work with and places that people talk about, several studies have been designed to discover individually meaningful places. Zhou et al. (2007) carried out an experiment to discover individually meaningful places using GPS trajectories. They conducted their experiment by asking the subjects to carry a GPS-enabled mobile device to record the GPS trajectories during their daily life. At the same time, a diary was also required from the subjects to log their meaningful places, which served as the ground truth to evaluate the algorithm's performance later. After a temporal preprocessing that eliminated GPS readings with speeds greater than 0, which removes the ones collected while driving, and the ones within a small distance of the previous reading, which reduces the amount of data, the DJ-Cluster algorithm (Zhou et al. 2004) was performed on the GPS data for place identification. This research proposes a general experimental framework on discovering individually meaningful places and its result indicates that clustering algorithm functions well on identifying individually meaningful places.

A thorough analysis on individual place identification was done by Nurmi (2009) in his PhD thesis. He introduced different location systems with their characteristics and the general process of place identification, mainly focus on GPS data. Detailed evaluation and comparison of existing place identification algorithm were also complied. He also pointed out that people's information needs of a place depend on their existing knowledge of the place.

In summary, modeling individual familiarity with places can help us to better understand people's knowledge of the environment, as studied in many disciplines like human geography and environmental psychology, as well as to provide location-based services adapted to people's prior spatial knowledge. However, this aspect has seldom been addressed in existing research. In the following, as a starting point, we propose an approach to rank individual familiarity with places using Foursquare check-in data.

3 Identification of Individually Meaningful Places: A Comparison of Existing Clustering Algorithms

In order to rank individual familiarity with places, it is necessary to identify individually meaningful places first. The check-in data of a user generally contain many check-in locations distributed in the user's activity space. Even though each check-in location has its own name and meaning in Foursquare system, it may share a same meaning with some other locations nearby to a certain individual. For example, some people check in different shops in a mall, but consider them all together as the shopping mall. Clustering algorithms can therefore be applied to find groups containing similar check-in locations within the data. Identified clusters are potential meaningful places to the user. Seeing the fact that there are many clustering algorithms already, a comparison of existing algorithms was conducted to choose an appropriate one.

3.1 Existing Clustering Algorithms

Based on different cluster models, clustering algorithms can be categorized in four main categories: hierarchical clustering methods, partitioning clustering methods, density based clustering methods and model-based clustering methods.

3.1.1 Hierarchical Clustering Methods

The main idea behind hierarchical clustering is that nearby objects are likely to be more similar than the objects far away (Heller and Ghahramani 2005). There are no input parameters needed, but a termination condition is usually essential to stop the merging or division. The most commonly used hierarchical clustering method is SLINK (Single-Link) (Sibson 1973).

3.1.2 Partitioning Clustering Methods

From an initial partitioning, partitioning clustering methods iteratively relocate data points between clusters until an optimal partition is attained (Äyrämö and Kärkkäinen 2006). The most popular algorithm of this kind is K-means, which requires a specified cluster number "K" and is sensitive to both the initial configuration and noise (MacQueen 1967).

3.1.3 Density-Based Clustering Methods

Density-based clustering algorithms tend to discover dense regions in a data space. The most extensively used one is DBSCAN (Density-Based Spatial Clustering of Applications with Noise) which is based on the idea that every point in a cluster should have a neighborhood of a given radius (given parameter: Eps) with at least a minimum number of points (given parameter: MinPts) in it. It has the ability to discover clusters with arbitrary shapes, insensitive to noise and the ordering of the points in the database (Martin Ester et al. 1996).

3.1.4 Model-Based Clustering Methods

Model-based clustering methods assume that the given data set is generated by a mathematical model and attempt to optimize the fit between them (Rokach and Maimon 2005). A widely used algorithm in this category is Expectation- Maximization Algorithm for Gaussian Mixture Model (EM algorithm for GMM) (Reynolds 2009). The EM Algorithm is a general approach to find maximum likelihood parameters by estimating the parameters iteratively between an expectation step and a maximization step (Dempster et al. 1977; Fraley and Raftery 2002).

3.2 Evaluation

An experiment was carried out to compare the four algorithms (SLINK, K-means, DBSCAN and EM Algorithm for GMM) introduced above.

3.2.1 Study Design

We ended up with 12 participants (9 females, 3 males) living in different cities in Asia and Europe. Their ages ranged from 23 to 50. The participants differed widely in their occupations, including financial sector, bio technology, education, health, information technology, logistic and social work.

Foursquare API was employed to collect the participants' check-in data. We retrieved their check-in histories through a webpage. Then we presented each participant's own check-in history to him/her and asked him/her to write down a list of meaningful places, together with the meaning of the places and the check-in locations included in each of the place. The meaning of a place refers to how that person understands and names the place, such as "home" and "university campus". This list was regarded as the ground truth to evaluate the performance of the algorithms later.

We then ran the four algorithms with identical parameter settings on the 12 data sets. For the K-means algorithm, we set "K" according to the cluster numbers suggested by the EM algorithm for the same data set. Based on empirical tests, the distance threshold of SLINK was set to 100 m and so as the Eps value in DBSCAN. Another parameter in DBSCAN, the MinPts, was given to 3.

To clarify the confusion between participants provided lists and algorithms' results, an interview with each participant was made at the end of the experiment. During the interview, issues like spurious places, which are the places identified by the algorithms and are actually sub-clusters of participants' provided meaningful places, but not on their lists, were discussed and their advices were gathered.

3.2.2 Evaluation Metrics

Precision and recall were used to measure the accuracy of a place identification algorithm. A "tolerance factor" was introduced to represent the spurious places. They were not correctly identified places but also meaningful to the participants, so we considered them in the tolerance factor. Utilized by Nurmi (2009) to evaluate the algorithms, F1-score, the harmonic mean of precision and recall, was also employed. The definition of precision, recall, tolerance factor and F1-score are:

$$\text{precision} = \frac{\text{correct}}{\text{NAC}}$$

$$\text{recall} = \frac{\text{correct}}{\text{NUC}}$$

$$\text{tolerance factor} = \frac{\text{spurious}}{\text{NAC}}$$

$$\text{F1} - \text{score} = \frac{2 \times \text{precision} \times \text{recall}}{\text{precision} + \text{recall}}$$

NAC in the above formulas refers to the number of algorithm generated clusters and NUC refers to the number of user provided clusters.

Table 1 Comparison of the algorithms

Algorithm	Precision	Recall	Tolerance factor	F1-score	Precision + tolerance factor
SLINK	0.516	**0.647**	0.281	**0.574**	0.797
K-means	**0.672**	0.441	0.134	0.533	0.806
DBSCAN	0.631	0.520	**0.286**	0.570	**0.917**
EM	0.537	0.353	0.119	0.426	0.656

The figures in bold are the highest in each category

3.3 Results and Discussions

Table 1 shows the overall performance of all the algorithms.

The results indicate that most algorithms do not have a good balance between precision and recall. K-means has the best precision but the second worst recall, while SLINK has the highest recall value and F1-score but the lowest precision value. Considering the sum value of the precision and the tolerance factor, which means the rate of meaningful places among all the places discovered by the algorithm, DBSCAN achieves the best performance. With a score of 0.917, nearly all the places discovered by it are actually meaningful to the participants. In general, EM algorithm for GMM has the poorest performance, while DBSCAN balances the best among the four algorithms. Accordingly, in our proposed method, DBSCAN is employed to identify individual meaningful places from Foursquare check-in data.

4 Ranking Individual Familiarity with Places

4.1 Methodology

After identifying individually meaningful places, we focus on ranking these places according to an individual's familiarity with them.

Suggested by Tuan (1977) and Mehrabian and Russell (1974), the influencing factors of the familiarity with a place consist of the frequency of visits as well as the extensity and intensity of the experiences there. The frequency of visits, which corresponds to the check-ins on Foursquare, has already been employed to discover individually meaningful places. Although the extensity of the experiences in a place can be represented by the duration of stay there, estimation of duration from Foursquare check-in data is impractical, due to the different check in habits of users (Lindqvist et al. 2011). However, the intensity of the experiences can be inferred by the tagging activities in that place.

Therefore, we turn to measure the intensity of the experiences. When an individual checks in at a place via Foursquare, three tagging activities often happen as well: "shout" (micro blog writing, i.e., adding a short text description to the check-in), "photos" (photo taking activity, i.e., adding photos to the check-in), and "like" (marking a check-in as "like"). All these activities increase the intensity of experience in that place. To measure the familiarity with a place based on the check-ins, a weighting is introduced based on these activities:

1. Each check-in has an initial weighting of 1;
2. If there is a "shout" in a check-in, the weighting will be incremented by 1;
3. If there are photos in a check-in, the weighting will be incremented by 1;
4. If the check-in is marked "like" by the user himself/herself, the weighting will be incremented by 1.

Therefore, the weighting of a check-in is at least 1 and can sum up to at most 4.

The weighting of each discovered place is then the sum of the weightings of all the check-ins belonging to this place. The familiarity with each place is represented by the weighting. Since the numerical values of the weightings do not have actual meanings in reality, we then rank the discovered places according to their weightings. The ranking of the discovered meaningful places stands for the familiarity of that individual with them. Thus, our method provides the ordinal measure of individual familiarity with places.

4.2 Evaluation

4.2.1 Study Design

To evaluate our method, a website was made to rank individual familiarity with places and an online survey was implemented. During the online survey, each participant was asked to log in to their Foursquare account. Once connected, his/her check-in history was then retrieved and based on which the meaningful places were identified using DBSCAN and ranked using the proposed method. These places were then randomized and listed on the screen. A participant could have lots of places discovered, for simplicity, if more than 10 places were discovered, only the top 10 places would be presented to him/her. Each participant was then asked to rank these places according to his/her familiarity with them (Fig. 1), which was considered as the ground truth to evaluate the performance of the proposed method.

In order to find out whether the proposed method is able to model individual familiarity with places, a random ranking of these places was also created for comparison. The reason of using a random ranking for comparison was mainly due to the fact that there was no other existing method for ranking individual familiarity with places to compare with. Taking the privacy issues into account, only participant's ranking, our calculated ranking and the random ranking were uploaded to our database.

In total, we received 23 effective responses from 14 females and 9 males. Their ages ranged from 23 to 50 with an average of 27. Students were the major group,

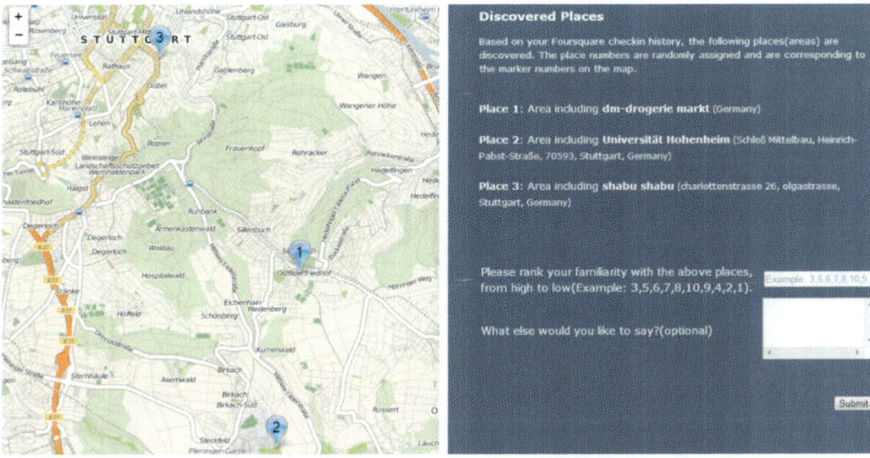

Fig. 1 Screenshot of the webpage where users can input their ranking of the places

while no retiree took part in. The occupations covered by the participants were financial sector, management, life and physical sciences, engineering, health, social service, education and transportation.

4.2.2 Evaluation Metrics

Spearman's rank correlation coefficient (Lehman et al. 2005), denoted by ρ, which measures the strength of association between two ranked variables, was employed to evaluate the correlation between the participant's ranking and our ranking, and the correlation between the participant's ranking and the random ranking.

In order to properly interpret the results of ρ to determine whether our ranking is significantly different with the random one, paired t-tests were then performed (Texasoft 2008).

4.3 Results and Discussion

Figure 2 shows the statistical description of the ρ value for each ranking. Figure 3 is a detailed comparison of ρ calculated for different rankings of each participant.

Comparing the mean values of ρ, the ranking generated by our method has a much stronger association with the participants' ranking than the random one. With a two-tail p-value smaller than 0.05 ($t(22) = 2.1189$, two-tail $p = 0.0456$), it is evidenced that the means of ρ_method is significantly larger than that of ρ_random. In other words, the correlation between the ranking generated by our method and the participants' ranking is significantly stronger than that between the random

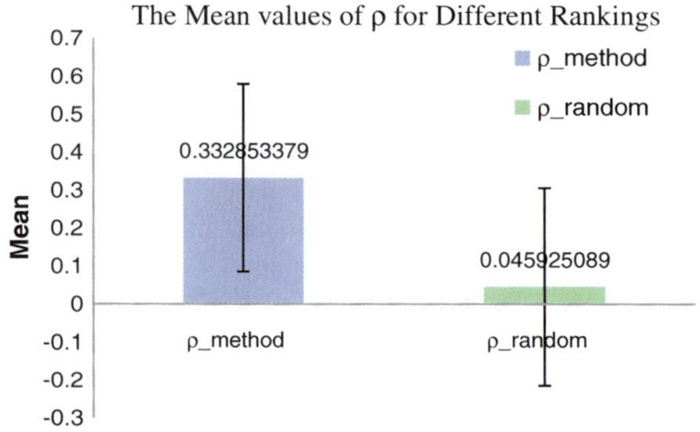

Fig. 2 The mean values of ρ for different rankings. "ρ_method" (*blue*, 0.333) stands for the ρ value calculated between the ranking of our proposed method and the participant's ranking. "ρ_random" (*green*, 0.046) is the ρ value calculated between the random ranking and the participant's ranking. Vertical error bars denote 95 % confidence intervals

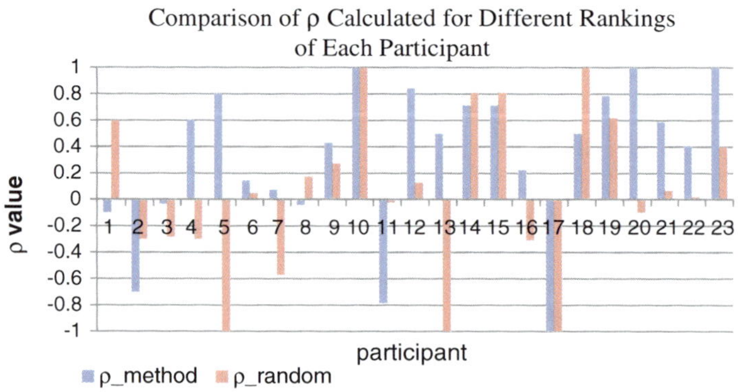

Fig. 3 Comparison of ρ calculated for different rankings of each participant

ranking and the participants' ranking. Suggested by the superiority over the random ranking, we conclude that our ranking method is able to rank individual familiarity with places.

The overall results imply the positive association between the ranking of our proposed method and participant's ranking, but to our surprise, from the figures, the association does not seem to be strong. The very simple weighting scheme may be responsible for it. A place is weighted according to the user's tagging activities at it. However, at very familiar places (e.g., home, office), users may not set any tagging activities beyond check-ins. As the frequency of visits is also in our concern and we assume that very familiar places are more frequently checked in, a more comprehensive weighting scheme should be introduced to emphasize the importance of the frequency. Also for the different tagging activities, their influences may not be the same. More detailed insights on them may help to place differentiated weightings. Even for the same tagging activity, for instance, micro blog writing, using different words express different intensities of the activity, so as the familiarity. In this case, employing natural language processing on the text descriptions may lead to a better outcome.

Another reason might be the limitation of Foursquare API, which only returns the most recent check-ins of an individual. Collecting Twitter messages that contain Foursquare check-ins might help to address this limitation (Noulas et al. 2011). In addition, starting from the connection between Twitter and Foursquare, more social media can be involved to rank individual familiarity with places, e.g., Facebook, Flickr and Twitter.

Some of the participants have commented on the survey. One of them mentioned that it is a bit puzzling to give an accurate ranking of the places, because some of them have similar importance to him. Thus, a categorization of individual importance might be preferable over a ranking in our future study.

5 Conclusion and Future Work

We put forward an approach to rank individual familiarity with places using location-based social media, particularly Foursquare check-in data. It firstly identifies individually meaningful places by clustering the check-ins. Then, it ranks the discovered meaningful places according to the weighting of each check-in belonging to that place. Based on the ranking, the individual familiarity with each place is generated.

Results of the evaluation show that, among four existing clustering algorithms (SLINK, K-means, DBSCAN and EM Algorithm for GMM), DBSCAN has the best performance in identifying individually meaningful places from Foursquare check-in data. More importantly, the evaluation shows that to some extent, our proposed method is able to rank individual familiarity with places.

Results of this study can be used to provide location-based services (e.g., mobile guides and navigation systems) that are adapted to users' priori spatial knowledge. It can also help to gain a better understanding of people's spatial knowledge and environmental perception.

To improve the approach and attain better results, more sample data could be collected so that a supervised method could be applied on it. A deeper insight on the influencing factors of human familiarity with places and detailed analyses on the text descriptions along with the check-ins can refine the weighting scheme. The knowledge fading effect could be considered by paying more attention on the time stamps of the check-ins. Other social media can also be replenished. A categorization of individual importance instead of a ranking could be considered in future studies.

Acknowledgments Wangshu Wang is supported by China Scholarship Council (CSC). Special thanks to Dr. Haosheng Huang and Prof. Georg Gartner for their supervision and useful comments to the draft of the paper.

References

Ashbrook D, Starner T (2002) Learning significant locations and predicting user movement with GPS. In: Proceedings of the 6th IEEE international symposium on wearable computers, pp 101–108

Äyrämö S, Kärkkäinen T (2006) Introduction to partitioning-based clustering methods with a robust example. Reports of the Department of Mathematics Information Technology, University of Jyväskylä (Series C. Software and Computational Engineering, 1/2006)

Chen WC, Battestini A, Gelfand N, Setlur V (2009) Visual summaries of popular landmarks from community photo collections. In: Proceedings of the 17th ACM international conference on multimedia, ACM, New York, NY, USA, pp 789–792

Cheng Z, Caverlee J, Lee K, Sui D (2011) Exploring millions of footprints in location sharing services. Presented at the proceedings of the fifth international conference on weblogs and social media, The AAAI Press, Barcelona, Catalonia, Spain

Cranshaw J et al (2012) The livehoods project: utilizing social media to understand the dynamics of a city. In: Proceeding of 6th international AAAI conference weblogs and social media, AAAI Press, pp 81–88

Dempster A, Laird N, Rubin D (1977) Maximum likelihood from incomplete data via the EM algorithm. J Roy Stat Soc 39(1):1–38

Edelsbrunner H, Kirkpatrick D, Seidel R (1983) On the shape of a set of points in the plane. IEEE Trans Inf Theor 29(4):551–559

Ester M et al (1996) A density-based algorithm for discovering clusters in large spatial databases with noise. In: Proceedings of second international conference on knowledge discovery and data mining, pp 226–231

Flickr (2008) http://code.flickr.net/2008/10/30/the-shape-of-alpha/

Fraley C, Raftery AE (2002) Model-based clustering, discriminant analysis, and density estimation. J Am Stat Assoc 97:611–631

Heller KA, Ghahramani Z (2005) Bayesian hierarchical clustering. In: Proceedings of the 22nd international conference on machine learning, pp 297–304

Hays J, Efros A (2008) IM2GPS: estimating geographic information from a single image. In: Presented at the IEEE conference on computer vision and pattern recognition, 2008, CVPR 2008, CVPR 2008, pp 1–8

Hollenstein L, Purves RS (2010) Exploring place through user-generated content: using Flickr tags to describe city cores. J Spatial Inf Sci 1:21–48

Kennedy L, Naaman M, Ahern S, Nair R, Rattenbury T (2007) How Flickr helps us make sense of the world: context and content in community-contributed media collections. In: Proceedings of conference on multimedia, ACM, New York, NY, USA, pp 631–640

Keßler C, Maué P, Heuer JT, Bartoschek T (2009) Bottom-up gazetteers: learning from the implicit semantics of geotags. In: Janowicz K, Raubal M, Levashkin S (eds) GeoSpatial semantics. Springer, Berlin, pp 83–102

Lehman A et al (2005) JMP for basic univariate and multivariate statistics: a step-by-step guide, SAS Press, Cary, NC, pp 123

Lindqvist J et al (2011) I'm the mayor of my house: examining why people use foursquare. In: Proceedings of the SIGCHI conference on human factors in computing systems, pp 2409–2418

MacQueen JB (1967) Some methods for classification and analysis of multivariate observations. In: Proceedings of 5th berkeley symposium on mathematical statistics and probability, vol 1, University of California Press, Berkeley, pp 281–297

Marmasse N, Schmandt C (2000) Location-aware information delivery with ComMotion. In: Proceedings of the 2nd international symposium on handheld and ubiquitous computing (HUC), vol 1927, Springer, Berlin, pp 361–370

Mehrabian A, Russell JA (1974) An approach to environmental psychology. MIT Press, Cambridge

Naaman M, Zhang AX, Brody S, Lotan G (2012) On the study of diurnal urban routines on twitter. In: Presented at the proceedings of the sixth international conference on weblogs and social media, The AAAI Press, Dublin, Ireland

Nurmi P (2009) Identifying meaningful places. PhD thesis, University of Helsinki, Helsinki, Finland

Noulas A, Scellato S, Mascolo C, Pontil M (2011) Exploiting semantic annotations for clustering geographic areas and users in location-based social networks. The social mobile web, volume WS-11-02 of AAAI workshops, AAAI

Rattenbury T, Naaman M (2009) Methods for extracting place semantics from flickr tags. ACM Trans Web 3(1):1

Reynolds DA (2009) Gaussian mixture models. encyclopedia of biometrics, pp 659–663

Rokach L, Maimon O (2005) Clustering methods. In: Rokach L, Maimon O (eds) Data mining and knowledge discovery handbook, pp 321–352

Shemyakin FN (1962) General problems of orientation in space and space representations. In Anan'yev BG et al (ed) Psychological science in the USSR, vol 1, NTIS Report No. TT6211083 (pp 184–255). Office of Technical Services, Washington, DC

Sibson R (1973) SLINK: an optimally efficient algorithm for the single-link cluster method. Comput J 16(1):30–34

Texasoft (2008) Understanding statistical hypothesis testing. URL http://www.stattutorials.com/understanding-hypothesis-testing.html

Tuan YF (1977) Space and place: the perspective of experience. University of Minnesota Press, Minneapolis, pp 138–184

Zheng Y, Li Y, Zha Z, Chua T (2011) Mining travel patterns from GPS-tagged photos. In: Lee K, Tsai W, Liao H, Chen T, Hsieh J, Tseng C (eds) Advances in multimedia modeling. Springer, Berlin, pp 262–272

Zhou C, Frankowski D, Ludford P, Shekhar S, Terveen L (2004) Discovering personal gazetteers: an interactive clustering approach. In: Proceedings of ACMGIS

Zhou C, Frankowski D, Ludford P, Shekhar S, Terveen L (2007) An experiment in discovering personally meaningful places: an interactive clustering approach. ACM Trans Inf Syst 25(3):12

Part IV
Innovative LBS Applications

A Space Time Alarm

Adrian C. Prelipcean, Falko Schmid and Takeshi Shirabe

Abstract Many modern mobile communication devices are equipped with a global positioning systems (GPS) receiver and a navigation tool. These devices are useful when a user seeks to reach a specified destination as soon as possible, but may not be so when he/she only needs to arrive at the destination in time and wants to focus on some activities on the way. To deal with this latter situation, a method and device called "Space Time Alarm" is presented for helping the user reach the destination by a specified deadline. It does so by continuously and efficiently computing how much more time the user may stay at his/her current location without failing to reach the destination by the deadline. Advantage of this approach is that it works completely in the background so that the user's en route activities will not be interfered with.

Keywords Alarm · Space time · Deadline · Route improvisation

1 Introduction

It is almost unthinkable but what would our modern life be like without clocks? Every morning we would wake up unsure if we are going to make the first shift of work. At a station we would wait for a next train without knowing when it will come.

A.C. Prelipcean · T. Shirabe (✉)
School of Architecture and Built Environment, Division of Geoinformatics,
KTH Royal Institute of Technology, Stockholm, Sweden
e-mail: shirabe@kth.se

A.C. Prelipcean
e-mail: acpr@kth.se

F. Schmid
Cognitive Systems Group, Universität Bremen, Bremen, Germany
e-mail: schmid@informatik.uni-bremen.de

G. Gartner and H. Huang (eds.), *Progress in Location-Based Services 2014*,
Lecture Notes in Geoinformation and Cartography,
DOI 10.1007/978-3-319-11879-6_13

Soon after lunch we would already start thinking about when we call it a day to go to pick up children from school. Actually, all these might be unnecessary worries because none of us (including our employers, train companies, and schools) would have the sense of punctuality. Of course, our circadian rhythms or environmental conditions could still tell us approximately where we are in the day. However, it would be impossible to schedule our daily activities as precisely as we do.

Here is a naïve question: will a time come when we wonder how we could live without knowing our exact location (not just time)? Maybe. With the advance of location detection technology such as global positioning systems (GPS) and mobile computing technology such as smartphones, we now experience, in our everyday life, a wide range of services—commonly known as "location-based services" (e.g., Spreitzer and Theimer 1994; Jiang and Xiaobai 2006; Raper et al. 2007 for overviews)—that are customized according to where we are. Location-based services can be even more sophisticated together with the consideration of time. For example, Raubal et al. (2004) has discussed a use of "time geography" (Hägerstrand 1975; Miller 2005) and "affordance theory" (Gibson 1977) to customize a sequence of activities subject to spatial and temporal constraints. While it may be too ambitious to fully optimize a schedule of all daily chores, but existing technology can already do a lot for us to make decisions concerning both location and time (see, e.g., Abdalla (2012) for his "Geo-Temporal Task Planning Application"). Nowadays, friends can exchange their location coordinates (instead of describing to each other where they are) through their smartphones, and use online mapping and spatial query services to find a coffee shop they can meet at the earliest possible time.

No matter how well a schedule is elaborated, it will be useless if it is missed. This is why the alarm function is a useful addition to clocks. Knowing that the alarm will go off when a specified time is passed, we can focus attention on the task at hand.

In the era when highly precise and accurate measurements of location and time are available to (almost) anyone anytime anywhere, what is a spatio-temporal counterpart to the alarm? To answer this, let us put ourselves in a somewhat familiar situation: we need to leave a hotel room for a university auditorium to give a talk at 10:00 am. We can still use the conventional alarm, but in this case, we need to do some calculation, because where the activity is performed is not where the alarm sounds. So we estimate how long it takes from the hotel to the university and back calculate what time we need to leave the hotel, and set the alarm accordingly. Here we can do the travel time estimation mentally or with assistance of mapping and/or navigation software. There are many online services that help us find a route and estimate its travel time.

Suppose further that the talk has been postponed to 2:00 pm, so that we have some time to explore downtown. Then, with the help of the navigation software, can we set the alarm to a right time to stop our exploration? This may not be as easy as in the previous case (where we are sitting in the hotel room) because we are moving around in the city, which means that it takes different amounts of time to get to the destination from different locations, which, in turn, means that the alarm must go off at different times at different locations. Still, if we do not mind to have the navigation software re-compute a route and travel time to the destination every time we change our location (perhaps by a specified distance or longer, as employed in Abdalla 2012). Unfortunately, this would lead to a dilemma: if the re-computation is done frequently, it would cost an exceedingly large amount of computing time/resource; otherwise, the travel time might not be computed in a timely manner. In any case, the alarm would not work timely or reliably. It might be an option to use a system proposed by Shirabe (2011), which continuously labels every alternative move at the current location with the estimated arrival time. However, this might impose too much cognitive load on the user as these labels must be updated rapidly and placed on a limited-size screen. As such, the idea of an alarm that takes into account location is interesting but not without limitations.

In view of the problem described above, we have posed a research question: Is there a computationally efficient (with respect to running time and storage space) approach to calculating how long one can remain at one's current location without failing to reach a specified destination by a specified deadline, and if so, how it can or should be utilized by a mobile user. This paper offers an answer by proposing a method and device, called "Space Time Alarm" (STA), for continuously tracking the user in space and time and alarming when the user has to leave the current (spatial) location in order to reach the destination by the deadline. Like the conventional alarm, it does this while running in the background without needing the user's attention. The remainder of the paper is organized as follows: Sect. 2 describes how the space time alarm works. Section 3 shows how a prototype of the alarm was implemented. Section 4 discusses the alarm's benefits, limitations, and possible extensions. Section 5 concludes the paper.

2 Methods

2.1 Assumptions and Data

One crucial step of STA is the estimation of a user's travel time. This is done based on two assumptions: (1) the user moves along streets, not through buildings or open fields and (2) the user moves at a constant speed. The first assumption implies that the user can turn from one street to another at a street intersection and also turn around anywhere on a street.

The street network is represented in digital form by a graph such that nodes represent street intersections and arcs represent street segments connecting two adjacent intersections. Each arc has an associated value representing the time of its traversal, which is obtained by dividing its geometric length by the user's speed. In this representation, if the user moves from node i to node j along arc(i, j), his/her location is specified by two pointers pointing at arc(i, j) and node i, and a numerical value indicating the travel time from node i to that location.

2.2 Components

STA consists of four general components:

- **data input/output means**, which accepts input from a user, and reads data from the storage means and communicate them to the user in various forms (e.g., visual, audio, and haptic),
- **data storage means**, which stores relevant data (including the digital street network and user input),
- **data processing means**, which reads data from the data storage means, processes them, and writes results to the data storage means, and
- **location and time detection means**, which detects the user's location and time.

2.3 Computational Steps

STA involves five computational steps, as presented in Fig. 1. The details of each step are described below.

2.3.1 Destination and Deadline Specification

The data input means accepts input specifying a destination and a deadline through the data input means. Like a conventional alarm, the deadline can be specified in the format of day-hour-minute or the like. Like an existing navigation system, the destination can be specified in the form of a postal address, a place name, or geographic coordinates or by pointing at its approximate location on a reference map. Any of these inputs is then converted to a node in the digital street network.

2.3.2 Shortest Travel Time Computation

The data processing means computes the shortest travel time to the destination from every node in the digital street network. It should be noted that this can be done by running an all-to-one shortest path algorithm (such as Dijkstra's algorithm, which was employed in our implementation as described in the next section) only once. Because relevant nodes are only those from which one can reach the destination by the deadline, the algorithm will be terminated when a node is labeled with a shortest travel time greater than the deadline minus the departure time.

Then the data processing means extracts only a subnetwork that spans the relevant nodes, and stores it in the data storage means. In the subsequent steps, this subnetwork is used instead of the whole network.

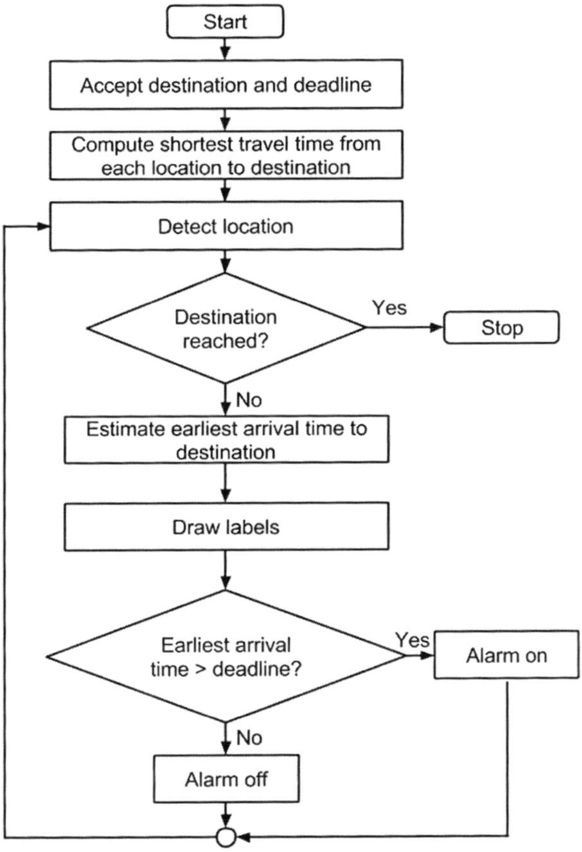

Fig. 1 Computational steps of space time alarm

2.3.3 Location Detection

The location and time detection means continuously receives a GPS signal (or any other location sensing data) containing coordinate data of the user's location. It is then converted to a point in the digital street network by an algorithm similar to curve-to-curve map matching. The algorithm takes into account an immediate history of the user's locations, as presented by White et al. (2000). From this history, the user's moving direction is determined, too. This process is outlined below with reference to Fig. 2.

A point P_1 is the first point detected by a GPS, and snapped to arc(i, j) because the arc is closest to P_1 and the distance between them does not exceed a given threshold. Then, a point P_2 is detected by the GPS, and snapped to arc(i, j) because the arc is closest to P_2 and the distance between them does not exceed the threshold and the movement direction P_1 to P_2 agrees to the arc's direction. Then, a point P_3 is detected by the GPS, but not snapped to its nearest arc(j, l) because the movement direction P_2 to P_3 (more or less vertical) would not agree to the arc's direction

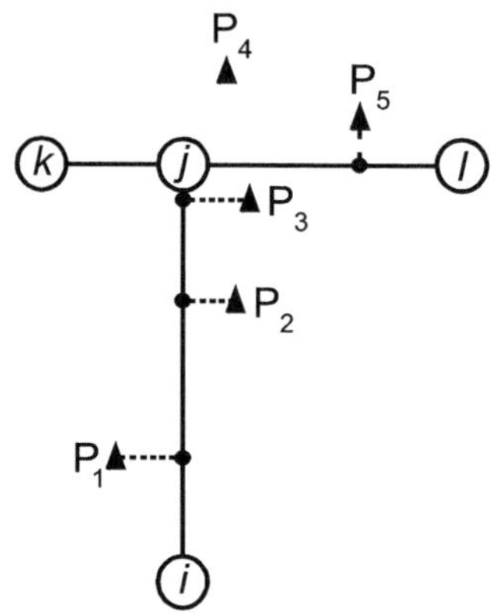

Fig. 2 Location detection. The circles represent nodes, the lines represent arcs, the triangles represent GPS points, and the dots represent network points determined by the matching algorithm

(more or less horizontal). Instead, P_3 is snapped to arc(i, j) because the distance between P_3 and the arc does not exceed the threshold and the movement directions P_2 to P_3 agrees to the arc's direction. Then, a point P_4 is detected by the GPS, but not snapped to its closest arc(j, l) because the distance between them exceeds the threshold. Finally, a point P_5 is detected by the GPS and snapped to arc(j, l) because the arc is closest to P_5, the distance between them does not exceed the threshold, and the movement direction P_4 to P_5 agrees to the arc's direction.

This location detection step will be revisited until the user reaches the destination.

2.3.4 Earliest Arrival Time Estimation

At every specified interval, the data processing means estimates the earliest arrival time at the destination by the following simple arithmetic. Suppose that the user is currently at a point p on arc(i, j) as shown in Fig. 3. Then, obviously, he/she must go through either node i or j to reach the destination. Thus, the earliest arrival time via node j is estimated as $t_c + t(p,j) + d(j)$, where t_c is the current time, $t(p,j)$ is the travel time from p to node j, and $d(i)$ is the shortest travel time from node j to the destination. Similarly, the earliest arrival time via node i is estimated as $t_c + t(p,i) + d(i)$. Therefore, the earliest arrival time, $t_e(p)$, from point p is given by:

$$t_e(p) = \min(t_c + t(p,j) + d(j), t_c + t(p,i) + d(i)) \quad (1)$$

.

It is important to note that Eq. (1) requires a trivial computation, as t_c is given by the location and time detection means at a constant interval, $d(i)$ and $d(j)$ have already been computed and stored in the data storage means, and $t(p,i)$ and $t(p,j)$ are linearly interpolated with point p, which has been already detected and stored in the data storage means, on arc(i, j).

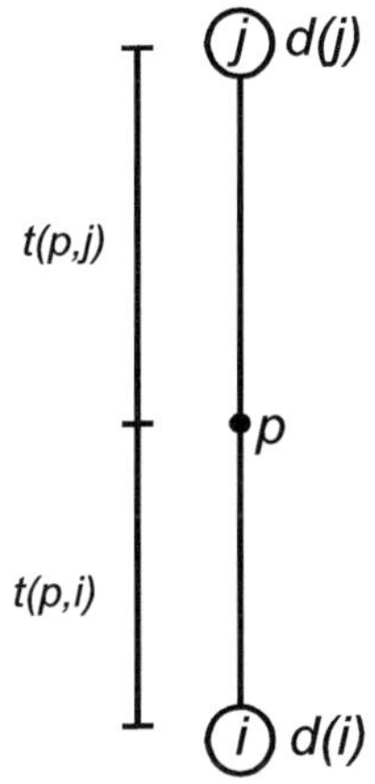

Fig. 3 Earliest arrival time estimation. From the current position p (represented by a *dot*) to the destination, it takes $t(p,i) + d(i)$ if node i (represented by a *circle*) is visited and *to* $t(p,j) + d(j)$ if node j (represented by a *circle*) is visited

2.3.5 Communication

The data processing means compares the earliest arrival time and the deadline. If the former is greater than the latter, the data input/output means generates an alarm signal (e.g., visual, audio, textual, or haptic) and communicates it to the user. Otherwise, it stops generating and communicating the alarm signal.

Optionally, some of the results stored in the data storage means may be also communicated to the user. For example, the earliest arrival time or the time left at the current location may be useful to the user.

3 Implementation

We have implemented STA based on a "full client architecture" (Jing et al. 1999) that connects each of multiple clients with limited storage and computing capability only once to a server with large storage (e.g., for street network data) and powerful computing capability (e.g., for shortest path computation). Having mobile devices as clients, this way allows us to serve multiple users at the same time. Figure 4 illustrates an overview of the architecture of the current implementation of STA.[1]

STA uses the main server of "OpenScienceMap (OSciM)" (Schmid et al. 2013), which plays roles of the data storage means and the data processing means. OSciM is a general platform to provide Android mobiles users with mapping and map rendering services. The OSciM server contains a PostgreSQL database management system (DBMS) with PostGIS extension, which stores the complete "OpenStreet-Map (OSM)" (Haklay and Weber 2008) data including worldwide street network

[1] A limited version that works only in Stockholm and Vienna is available at http://people.kth.se/~shirabe/SpaceTimeAlarmClock/STAC.html.

data and supports the management, query, and analysis of these data. The OSciM server uses an Apache Tomcat web server and implements several Java servlets to communicate with multiple clients efficiently. The PostgreSQL DBMS further contains the pgRouting extension, which specializes in geospatial routing functionality including Dijkstra's shortest path algorithm.

Each client of STA is an Android smartphone or any other Android device, which has a storage medium containing a SQLite DBMS, a central processing unit (CPU), a GPS receiver, and a graphical user interface (GUI), and is able to communicate with the OSciM server. The SQLite DBMS plays a role of the data storage means. The CPU plays the role of the data processing mean. The GPS receiver and the CPU clock serve as the location and time detection means. The Android GUI serves as the data input/output means.

With reference to Fig. 4, the workflow of STA is explained below. First, through the Android GUI (Fig. 5) the user specifies a destination by tapping on the corresponding location on a reference map or typing its address, place name or geographic coordinates in the textbox on the top of the display. This brings the user to the next dialog box in which the date/hour/minute of a deadline is to be selected (arrow 1). Then, a request for shortest travel time computation, together with the specified destination and deadline, is sent to OSciM server (arrow 2). In response to the request, the OSciM server runs a servlet requesting the PostgreSQL DBMS to run a shortest path algorithm and extract a subnetwork that spans all nodes from which the destination can be reached by the deadline. The subnetwork is then sent to the Android client using another servlet (arrow 3). The client stores it in the SQLite database and will not contact the server again.

The Android client then builds a quadtree indexing structure (Samet 1984) for all the nodes in the subnetwork in order to efficiently perform necessary spatial queries (e.g., retrieval of the node closest to a specified point or retrieval of the node having a specified ID) in the remaining steps. Then, the client's GPS receiver starts tracking the user's location every second. This frequency was chosen as a

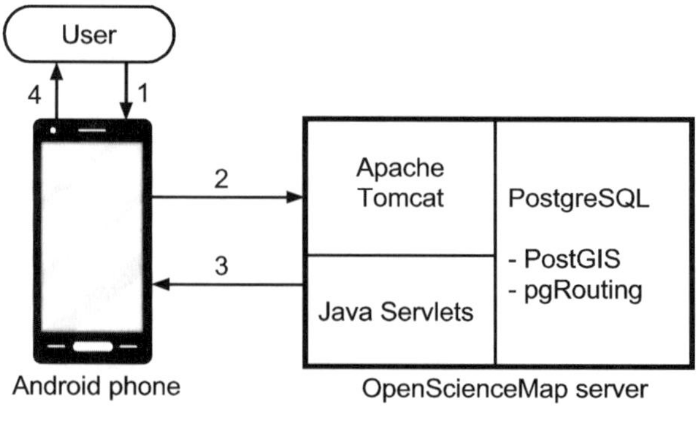

Fig. 4 Architecture of an implementation of STA. The OpenScienceMap server (*right box*) works as the STA server, and an Android device as a STA client (*phone image*) with which a user (*top oval*) interacts

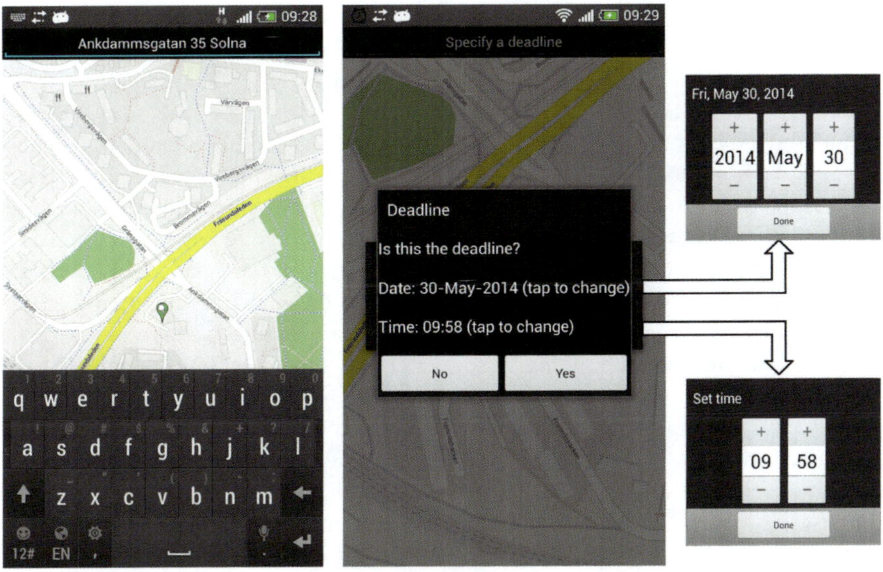

Fig. 5 Specification of a destination and a deadline in the STA research prototype. On the *left*, the map marker represents the location of a specified destination (which, in the present example, corresponds to the address typed in the *textbox* on *top*). In the *middle*, a specified deadline (which is, in the present example, set by default to 30 min from the current time) is shown. The date and time of the deadline can be modified through two dialog boxes shown on the *right*

compromise between the client's battery life and the accuracy in map matching. Each GPS point is then map-matched to a point in the subnetwork. Only the latest four map-matched points are stored in the SQLite database and used to reduce the uncertainty (due to GPS measurement noise) in the user's location and movement direction. Then, the earliest arrival time is estimated using Eq. 1 every second. If the estimated earliest arrival time is later than the deadline, the Android client sounds an alarm sound (arrow 4).

4 Discussion

In this section, we discuss several notable aspects of STA based on our experience with the present implementation of STA.

First of all, STA will not replace existing navigation systems but complement them, as its major purpose is limited to notifying the user of a right time to leave. To see this, imagine a tourist visiting an unfamiliar city. He/she initially uses STA and enjoys sites without distractions, and once STA goes off, he/she switches to a navigation system to take a fastest route and reach his/her eventual destination in time.

Second, although travel time computation relies on the assumption that the user moves at a constant speed (by default 1.0 m/s), this does not mean that STA fails if the user moves at a different speed. If the user slows down (or even stops), he/she loses time, i.e., STA will go off earlier. If the user moves faster (towards the destination), on the other hand, he/she gains time. These have been experienced during our initial test, which is exemplified by the following scenario. In response to our specification of a destination and a deadline, STA calculated that the alarm would go off in 10 min. Then we ran for a while and found this number increased to 12 min. Then, as we slowed down, the number stopped increasing and eventually started decreasing. Also, we intentionally went to the opposite direction from the destination, and the number decreased. Another interesting observation was that when we went through a building or across a park, we gained time because it had a similar effect as moving faster.

Third, if it is implemented in a full-client architecture, STA works offline as soon as it completes the step of shortest travel time computation (see Fig. 1). This is because a relevant subset of the digital street network (with each node having its shortest travel time to the destination) is thereafter stored in the client device. A positive consequence is that STA may be used without exposing the client's location (and the user's privacy) to the server.

Fourth, as long as GPS signals are available, STA works without interruptions or delays. Notice that the shortest travel time computation mentioned above is the most computationally expensive step, and it is executed only once when the destination is set. Other steps (such as location detection, estimation of the earliest arrival time and its comparison with the deadline) are done in real time but do not require much computing time or resource.

Finally, the main target users of STA are pedestrians having mobile computing devices such as smartphones. In this mode of transportation (i.e., walking), it is fairly safe to assume that the attribute and topology of the underlying mobility network are static, which means that the shortest travel times initially calculated remain valid until the destination is reached. For STA to accommodate other modes of transportation, however, some computational difficulties must be overcome. In the case of driving, the traffic conditions may change rapidly depending on the time of a day. Every time the digital street network is updated, STA needs to retrieve it from the data server and re-apply the shortest path algorithm to it. If such updates occur too often, STA will fail to alert in a timely manner. In the case of public transportation (e.g., bus and train), STA is expected to work fairly well, as the shortest travel time computation can be done according to published time tables, which are usually reliable. However, the corresponding network data still need to be updated if some unexpected events (e.g., traffic accidents and mechanical troubles) occur. Therefore, the success of the extension of STA depends on how well the shortest travel time (re)computation can be done with real-time network data.

5 Conclusions

We have presented a new location-based service, called Space Time Alarm (STA), for continuously monitoring the location of a user and alerting when the user has to leave the current location in order to arrive at a specified destination by a specified deadline. To do so, assuming that the user moves at a constant speed through a given street network, STA performs a simple logic comprising: (1) accepting a specification of a destination and a deadline and computing the shortest travel time from each node of the network to the destination, (2) tracking the user's location in the network in real time, (3) adding the current time and the travel time from the current location to the destination, and (4) generating an alarm signal if the result exceeds the deadline. Optionally it also shows the amount of time remaining at the current location before it becomes impossible to reach the destination in time. STA can be implemented as a stand-alone mobile application or integrated to an existing navigation system.

STA is expected to benefit any user who has a sufficient amount of time before going to his/her final destination and wants to utilize this spare time to engage in other activities (e.g., shopping and exploration) that require active attention to (and interaction with) the real environment. Once the destination and deadline are set, STA works in the background "calmly" (Weiser and Brown 1996) and lets the user enjoy the en route activities without the fear of being late. When the alarm goes off, the user is urged to head for the destination, with assistance of a navigation system if necessary.

The present implementation of STA is designed for pedestrians only. It may be tempted to extend it to other modes of transportations including driving and public transportation. Care must be taken, however, because this would bring about a greater degree of uncertainty and dynamics in arrival time estimation, which, in turn, would require not just one, but more likely frequent, update of network data and execution of a shortest path algorithm, which would cause a significant amount of computation and thus communication delay especially if STA is embedded in a mobile device.

Finally, an important implication of STA is that while location-based technology may continue to advance in terms of the capability of prescribing spatio-temporal decisions or plans (e.g., route directions) that optimize multiple, possibly conflicting, objectives and constraints, it can also offer less proactive assistance, which aims to softly influence one's behavior by limiting its intervention to the provision of warnings or suggestions only when undesired outcomes (e.g., being late) are anticipated to happen. The question is not about which of the two capabilities is more important but how to balance them according to the purpose and context and, if necessary, alternate them seamlessly.

Acknowledgments This project is partially supported by the German Research Foundation (DFG) through the Collaborate Research Center SFB/TR 8, the European Union Seventh Framework Programme—Marie Curie Actions, Initial Training Network GEOCROWD (grant agreement No. FP7-PEOPLE-2010-ITN-264994).

References

Abdalla A (2012) LatYourLife: a geo-temporal task planning application. In: Gartner G, Ortag F (eds) Advances in location-based services. Springer, Berlin, pp 305–325. doi:10.1007/978-3-642-24198-7_20

Android. http://www.android.com/. Last accessed on 08/17/2014

Apache Tomcat. http://tomcat.apache.org/. Last accessed on 08/17/2014

Gibson JJ (1977) The theory of affordances. In Perceiving, acting, and knowing, pp 67—82. ISBN 0-470-99014-7

Hägerstrand T (1975) Space, time and human conditions. In: Dynamic allocation of urban space , Lexington Books, Lexington, MA, pp 3–14. ISBN 0-347-01052-0

Haklay M, Weber P (2008) Openstreetmap: user-generated street maps. IEEE Pervasive Comput 7 (4):12–18. doi:10.1109/MPRV.2008.80

Jiang B, Xiaobai Y (2006) Location-based services and GIS in perspective. Comput Environ Urban Syst 30(6):712–725. doi:10.1016/j.compenvurbsys.2006.02.003

Jing J, Helal AS, Elmagarmid A (1999) Client-Server computing in mobile environments. ACM Comput Surv 31(2):117–157. doi:10.1145/319806.319814

Miller HJ (2005) A measurement theory for time geography. Geog Anal 37(1):17–45. doi:10.1.1.65.4624

PostGIS. http://postgis.net/. Last accessed on 08/17/2014

PostreSQL. http://www.postgresql.org/. Last accessed on 08/17/2014

pgRouting project. http://pgrouting.org/. Last accessed on 08/17/2014

Raper J, Gartner G, Karimi H, Rizos Chris (2007) A critical evaluation of location based services and their potential. J Location Based Serv 1(1):5–45. doi:10.1080/17489720701584069

Raubal M, Miller HJ, Bridwell S (2004) User□centered time geography for location□based services. Geogr Ann: Ser B Hum Geogr 86(4):245–265. doi:10.1.1.196.1642

Samet H (1984) The quadtree and related hierarchical data structures. ACM Comput. Surv (CSUR) 16(2):187–260. doi:10.1145/356924.356930

Schmid F, Janetzek H, Wladysiak M, Bo H (2013) OpenScienceMap: open and free vector maps for low bandwidth applications. In Proceedings of the 3rd ACM symposium on computing for development, ACM, p. 51. doi:10.1145/2442882.2442939

Shirabe T (2011) Information on the consequence of a move and its use for route improvisation support. In Spatial information theory, Springer, Berlin, pp 57–72. doi:10.1007/978-3-642-23196-4_4

Spreitzer M, Theimer M (1994) Providing location information in a ubiquitous computing environment (panel session). In ACM, 27(5):70–283. doi:10.1145/173668.168641

SQLite. http://www.sqlite.org/ Last accessed on 08/17/2014

Weiser M, Brown JS (1996) Designing calm technology. PowerGrid J 1(1):75–85. doi:10.1.1.123.8091

White CE, Bernstein D, Kornhauser AL (2000) Some map matching algorithms for personal navigation assistants. Transp Res Part C: Emerg Technol 8(1):91–108. doi:10.1016/S0968-090X(00)00026-7

Urban Emotions—Geo-Semantic Emotion Extraction from Technical Sensors, Human Sensors and Crowdsourced Data

Bernd Resch, Anja Summa, Günther Sagl, Peter Zeile and Jan-Philipp Exner

Abstract How people in the city perceive their surroundings depends on a variety of dynamic and static context factors such as road traffic, the feeling of safety, urban architecture, etc. Such subjective and context-dependent perceptions can trigger different emotions, which enable additional insights into the spatial and temporal configuration of urban structures. This paper presents the Urban Emotions concept that proposes a human-centred approach for extracting contextual emotional information from human and technical sensors. The methodology proposed in this paper consists of four steps: (1) detecting emotions using wristband sensors, (2) "ground-truthing" these measurements using a People as Sensors location-based service, (3) extracting emotion information from crowdsourced data like Twitter, and (4) correlating the measured and extracted emotions. Finally, the emotion information is mapped and fed back into urban planning for decision support and for evaluating ongoing planning processes.

Keywords People as sensors · Urban planning · VGI · Crowdsourcing · Emotion detection

B. Resch (✉) · G. Sagl
Department of Geoinformatics, University of Salzburg, Salzburg, Austria
e-mail: bernd.resch@geog.uni-heidelberg.de; bernd.resch@sbg.ac.at

B. Resch · A. Summa · G. Sagl
Institute of Geography—GIScience, Heidelberg University, Heidelberg, Germany

B. Resch
Center for Geographic Analysis, Harvard University, Cambridge, USA

P. Zeile · J.-P. Exner
Computergestützte Planungs- und Entwurfsmethoden (CPE),
University of Kaiserslautern, Kaiserslautern, Germany

G. Gartner and H. Huang (eds.), *Progress in Location-Based Services 2014*,
Lecture Notes in Geoinformation and Cartography,
DOI 10.1007/978-3-319-11879-6_14

1 Introduction

The development of a digital city into an intelligent city offers new opportunities to capture spatial and temporal data in near real time. The enabling driver is a continuous connection between the physical and the digital worlds by a variety of "sensors"—from calibrated measurement equipment to human sensors (Goodchild 2007). Recent developments use psycho-physiological measurements in urban space, for instance, to map emotions (Zeile et al. 2009), or mobile phone data and social network data to assess collective human behaviour patterns (Sagl et al. 2012). These new data and information layers can provide additional insights into the development of both the physical and social structures of inherently complex and dynamic urban environments.

Yet, the city as a functional construct shall not only be seen as a place of technological infrastructure, financial transactions, a network of technical nodes, a geographical agglomeration area or a political landscape, but more as an actuated multi-dimensional conglomerate of heterogeneous processes, in which the citizens are the central component (Resch et al. 2012). This interaction between humans and urban space, i.e., where, when and in particular how people respond to urban processes, needs more attention from a quantitative point of view in order to derive more reliable results compared to current urban analysis approaches.

Urban Emotions aims to address this shortcoming by providing a human-centred approach for extracting contextual emotion information from technical sensor data (measurements from calibrated bio-sensors) and human sensor data (subjective observations by citizens). The results are used in the domain of urban planning for decision support and the evaluation of ongoing planning processes (Zeile et al. 2014). Like this, the realization of a Smart City is not only to be tackled from a technological viewpoint (as most previous research efforts did), but from a human-centred viewpoint that claims that a city requires "Smart Citizens" to be intelligent itself.

Figure 1 illustrates the general *Urban Emotions* concept that comprises four steps: (1) detecting emotions using wristband sensors, (2) "ground-truthing" these measurements using a smartphone-based People as Sensors (Resch 2013) location-based service (LBS) in near real time, (3) extracting emotion information from crowdsourced data like Twitter (detecting the type of emotion), and (4) correlating the measured and extracted emotions. Subsequently, the emotion information is mapped and fed back into urban planning processes. The paper at hand mainly covers the two modules of ground-truthing emotion information using the People as Sensors concept and the extraction of emotion information from text-based Volunteered Geographic Information (VGI). For this reason, we employ a graph-based semi-supervised learning (SSL) algorithm, i.e., we use a small set of labelled training data to assign a label to each Tweet—where the process of *labelling* stands for assigning an emotional category to each Tweet (Sect. 4). Thus in contrast to previous approaches, which have relied on methods from a single discipline like GIScience, computational linguistics (CL), sociology, or computer science (CS), we propose a trans-disciplinary method.

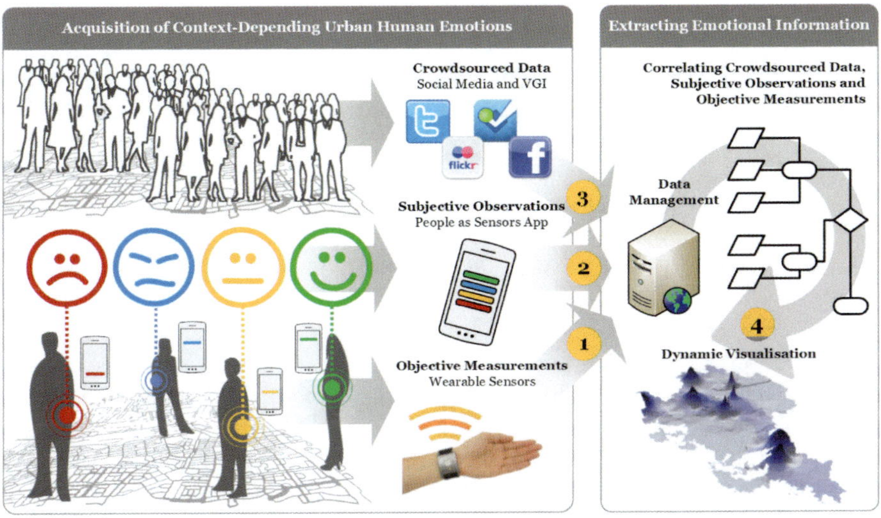

Fig. 1 Urban emotions concept: (*1*) emotion sensing, (*2*) ground-truthing using people as sensors, (*3*) extraction of emotion information from VGI, (*4*) correlating measured and extracted emotions; plus visualisation and enrichment of urban planning processes

This paper is structured as follows: After this introduction, we provide a description of related work including a clear identification of research gaps (Sect. 2). Then, we present our methodology for ground-truthing emotion measurements using a People as Sensors app (Sect. 3), followed by a description of our proposed methodology for extracting emotion information from Twitter Tweets (Sect. 4). Finally, we discuss our approach's integrability into urban planning processes (Sect. 5), and end the paper with a number of key conclusions and future research avenues (Sect. 6).

2 Related Work

For the scope of this paper, related work needs to be examined in three areas: using emotion measurements for urban applications (Sect. 2.1), emotion extraction from VGI (Sect. 2.2), and using emotion information in the field of urban planning (Sect. 2.3).

2.1 Measuring Emotions in the Urban Context

Salesses et al. (2013) describe an approach to extract information about citizens' perception of safety. They developed an online platform that allows people to compare and rate two randomly selected pictures showing different urban environments from a street-view perspective with respect to safety. This allows for a qualitative subjective assessment of a static situation (i.e., a picture). Yet, a continuous subjective perception and quantitative assessment with respect to the dynamic situational urban context, providing insights beyond a snapshot, is not addressed in the approach.

Gartner (2010) investigates the use of emotions to support way-finding tasks. The paper describes methods how emotions can be sensed and presents a conceptual framework for the use of emotion information in way-finding. However, no concrete field tests have been carried out, and no implementation and validation of the concept have been performed. The Urban Emotions approach goes one step further by applying emotion information in concrete real-world urban management and planning use cases. In addition, our approach performs a "ground-truth" for the emotional spike (Sect. 3).

The approach by Klettner and Schmidt (2012) aims to cartographically visualise emotions. In other words, the use of emotion information is addressed from a purely cartographic presentation viewpoint, while no in-depth analysis and feedback to real-world processes is done.

2.2 Extracting Emotion Geo-Information from VGI

The research field of *Sentiment Analysis* only deals with a word's, sentence's, or document's polarity, i.e., whether it conveys a positive, negative, or neutral sentiment. Additionally, research has been done to determine the expressed sentiment's strength (Liu and Zhang 2012). In the Urban Emotions approach, we use a more sophisticated emotion model as knowledge purely concerning a Tweet's polarity is not sufficient.

Emotion Detection from Tweets focuses on classifying Twitter posts according to a number of distinct emotions. The two approaches by Roberts et al. (2012) and Bollen et al. (2011) analyse the results of large events that cover the entire USA and influence Twitter traffic for one or several days. In doing so, singular small-scale variations of Twitter "traffic" might be overseen. For urban planning, these smaller events may be important as they affect smaller, local areas. Besides, both approaches lack the geographic component, which is essential to our approach as laid out in Sect. 1. Also, these previous efforts neglect the possibilities that arise for emotion detection from emoticon analysis. Finally, both approaches propose insufficient methodologies that we aim to improve: Roberts et al. (2012) only work in a supervised manner and thus require a large dataset to produce satisfying results. Bollen et al. (2011) do not evaluate their approach against a ground-truthed gold standard, but correlate their results manually to events that occurred at the same time the Tweets were sent. In our work, we overcome these methodological problems through using a semi-supervised learning approach, which can be applied to a dataset with few labelled, but numerous unlabelled instances (a smaller data set), and evaluated on a test set.

Another approach by Hauthal (2013) aims to detect emotions in VGI and to map emotional hot spots in a city. However, the approach works on a simple syntactical word-matching algorithm that is not able to cope with the complexity of unstructured text data like Twitter Tweets.

2.3 Using Emotion Information in Urban Planning

In the context of urban planning, only few research efforts deal with the question, how planners can make use of subjective feelings in urban surroundings and perceptions of citizens. Lynch's approaches of the "Image of the City" and "Mental Maps" suggest that citizens investigate a city during a walk and afterwards sketch a map of the investigated area out of their mind. Yet in this approach, test persons need excellent drawing skills to produce the maps (Sorin Matei et al. 2001). introduced a first digital approach of showing feelings in a digital map. This "Mental Maps Concept" visualises feelings not only on a digital map, but also on a three-dimensional model of the city for a better understanding of "geolocation of fear" in Los Angeles, USA. The art project "Biomapping" by Christian Nold was the first work that combined "emotional data" (physiological parameters like skin resistance levels) with GPS datasets (Nold 2009).

Despite such technical challenges, all experts in this field pointed out that citizens are the main and most important actor in urban planning processes. Thus, our approach focuses on using emotion information from a variety of sources and on feeding it back directly to urban planning.

3 Ground-truthing Emotion Measurements: People as Sensors

As laid out in Sect. 2, one of the major shortcomings of previous approaches is that emotions measured by technical sensors cannot be unambiguously correlated with a person's actual emotion (the type of emotion) and the cause why an emotion occurred (the context of the emotion). This is because currently available emotion sensors are only capable to detect a person's emotional spike, but not their causal trigger. Most current sensors observe a person's "additional heart rate", skin conductance, body temperature—the latter two variables are counter-rotating, i.e., in case an emotional spike occurs, skin conductance rises and body temperature drops because the person emits cold sweat.

To account for this shortcoming, we designed and implemented a People as Sensors LBS, through which persons can enter the emotional category and the according context in case an emotional spike occurs. In the design of the app, we aimed to fulfil common guidelines for mobile applications. First, we followed the design principles: "make it direct", "keep it lightweight", "stay on the page", "provide an invitation", "use transitions", and "react immediately", as defined by Scott and Neil (2009). According to these principles, we decided to design a simple interface that offers users the possibility to first input the type of emotion, followed by a screen to input the context of the emotion. Both inputs immediately guide the user back to the start page. Thus, the interface is lightweight in terms of design elements, the number of "clicks" required, and the information provided to the user. To comply with the requirement of displaying an invitation, we notify the user when their input is required; i.e., when the wristband sensor measures an emotional spike, the user is requested to input the type and context of the emotion. Furthermore, we decided to greatly avoid strong colouring of the app and assigning colours to emotions in order to prevent emotional biases (Mohammad 2011). Figure 2 shows the application interface in three steps, (1) the input for the type of emotion, (2) the input for the context, and (3) the main page for submitting the data.

In order to comply with the requirement of immediate feedback to the user, we intend to show a map containing the emotions of a city right after a user has entered their information. Like this, the motivation for entering emotion information shall be kept high and users can immediately "compare" their own impressions with other persons' perceptions.

In the next design phase, we also aim to improve the input modalities. One option that we are currently investigating is the use of a "colour wheel" (comparable to the RGB colour wheel) to enable the input of a combined emotion in a quasi-continuous emotion space instead of just a single emotion. Even though this does not comply with the idea of avoiding bias (s. above), it could be important to

Fig. 2 Three screens for entering the type of emotion, the context for the emotion and for sending off the information (*left to right*)

allow for inputting several emotions as emotional categories are mostly related to others and they do not appear as single isolated emotions. Furthermore, we investigate the possibility of allowing users to input the intensity of their emotion, which can be used for assessing the intensity's correlation with the spikes' amplitudes measured by the emotion sensor.

4 Emotion Extraction from VGI

As mentioned in Sect. 1, previous approaches of extracting information, particularly emotion information, from crowdsourced data have relied on methods from a single discipline. The approach proposed in this paper, to our knowledge, is the first method that extracts information in a trans-disciplinary algorithm, using methods from CL and GIScience.

Previously, little research has been conducted on emotion extraction on Twitter. Consequently, few emotion-annotated Tweet datasets are available (e.g., Roberts et al. 2012). Yet, none of these corpora is sufficient for our purpose because they all lack the geographical component. Thus, our workflow involves the creation of a manually annotated emotion-tagged Twitter corpus, which will be employed for both training and evaluation purposes.

4.1 Basic Workflow

The basic workflow for our approach is depicted in Fig. 3. It starts with a set of unlabelled, raw Twitter posts along with their metadata (location, time, user ID, post ID, etc.), and results in a set of labelled Tweets (where the labels are the emotion categories). The workflow comprises five distinct steps that are described below.

The **first step** is to filter the Tweets by language because in our first use case, we use only English Tweets to respect our annotators' language skills and facilitate manual inspection. Furthermore, the reduction to English language simplifies data processing in our initial field tests due to a large number of Natural Language Processing (NLP) resources.

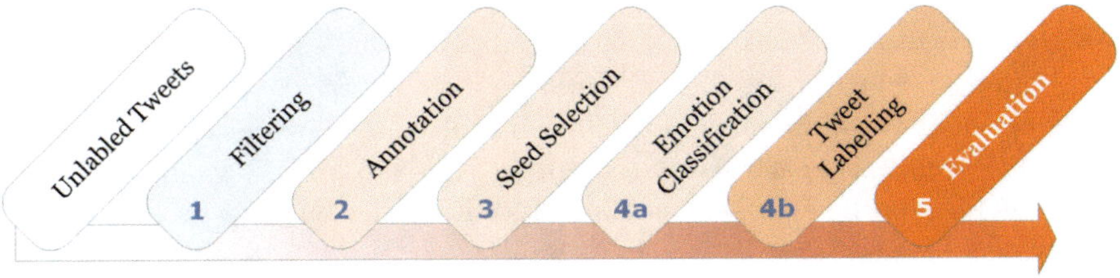

Fig. 3 Workflow for extracting emotion information from twitter tweets

The **second step** is the annotation process, in which Tweets are classified manually by volunteers with respect to whether they contain one of the emotions of interest or are considered neutral. Like this, we create a "gold standard" for the evaluation later in the workflow. This gold standard is the basis for a quantitative evaluation of our results to prove the effectiveness of our approach. In other words, a quantitative evaluation of our results is necessary to prove that our approach can actually detect a particular emotion in a single Tweet. As mentioned in Sect. 2, qualitative evaluation methods like manually correlating emotion detection results to large-scale events do not ensure this. For our research, we use the emotion model proposed by Ekman and Friesen (1971), which distinguishes between the six basic emotions joy, anger, fear, sadness, surprise, and disgust. The reasons for using this model are (1) that it consists a solid, well-established emotion model, and (2) that it is used in similar research, e.g., by Roberts et al. (2012), which ensures comparability of the datasets, and the results and proves that this theoretical model is applicable to Tweets.

The **third step** serves for selecting the subset of labelled Tweets that are employed as seeds for the subsequent graph-based semi-supervised learning (SSL) algorithm. Seed selection is a critical step in semi-supervised settings, as the seeds strongly influence the program's output as the program can only use labels that are contained in the seed set, which thus needs to contain all labels it is supposed to use. Additionally, the distribution of labels over the seeds influences the distribution of labels in the result. Thus, we are investigating two ways of seed selection, (1) to align the seed distribution with the actual emotion distribution in the entire dataset, or (2) to select uniformly distributed seeds.

The **fourth step** performs the actual emotion detection and classification, and is further divided into four sub-steps. Since this is the core part of our workflow, these steps are described in detail in Sect. 4.2 below.

The **fifth step** is the evaluation of the labelling process' results (Sect. 4.2) against our hand-annotated gold standard. The subjective nature of the task of identifying emotions necessitates the comparison of the computer's performance against that of human beings. A suitable evaluation measure must be applied, depending on the importance of precision ("positive predictive value") and recall ("sensitivity") or their ratio ("F1 score").

4.2 Similarity Computing and Label Propagation

As mentioned above, step four in our overall workflow is the central part of detecting emotion information. In this step, unlabelled Tweets are converted to labelled Tweets, i.e., each Tweet is assigned an emotional category. Figure 4 illustrates the workflow for the labelling procedure. The single steps are explained below.

First, the Tweet text is **pre-processed** using basic text processing algorithms. This process includes enriching the text with different kinds of semantic

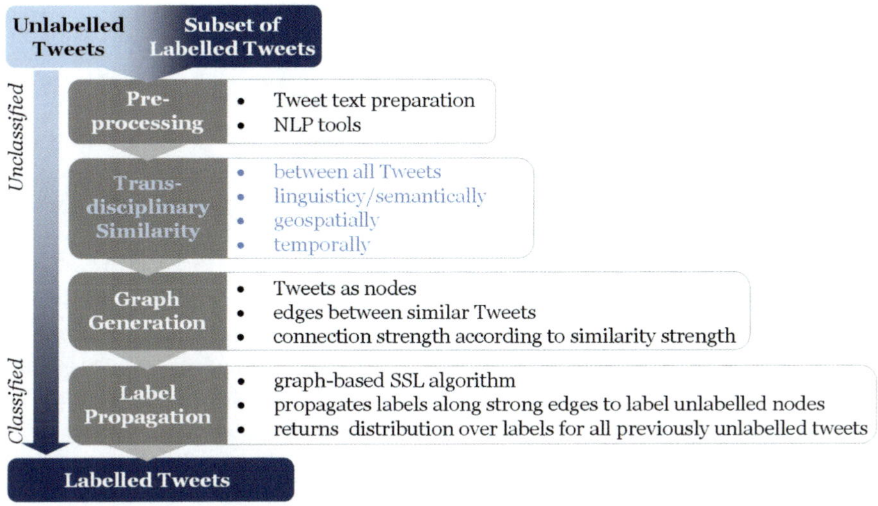

Fig. 4 Workflow for labelling Tweets in a trans-disciplinary approach to similarity

information to reveal the text's hidden structure. We apply the part-of-speech (POS) tagger presented by Owoputi et al. (2013) to detect emoticons, negations, and other specific POS. Pre-processing in our case also includes constraining possible emotion labels for unlabelled Tweets by comparing their content to the Affective Norms for English Words (ANEW) word list (Bradley and Lang 1999).

Second, the **similarity** between all Tweets, both seeds and unlabelled, is computed. This, along with the Tweets themselves, constitutes the input for the graph constructed in the next sub-step. We compute similarity according to three factors: *linguistic and semantic content, temporal distance,* and *geo-spatial distance.* We argue that Tweets, which are alike with respect to certain aspects, such as having emotional words, emoticons, and other linguistic features in common, will likely also carry the same emotion.

The main novelty of our approach is the combination of the concepts of similarity, distance and clusters that exist in both the domains of GIScience and computational linguistics, into a single model. Concretely in accordance with Takhteyev et al. (2012), we apply Waldo Tobler's First Law of Geography (Tobler 1970) to our research by assigning a higher similarity score to Tweets, which are closer in geographic space. In consequence, our combined model considers Tweets to be "similar" if the distance to each other is small in semantic space, in geographic space and in the time domain.

As for the CL dimension of the method, different degrees of similarity can be defined (Agirre et al. 2012): *completely equivalent*, as they mean the same thing; *mostly equivalent*, but some minor details differ; *roughly equivalent*, but some important information differs; *not equivalent*, but share some details; *not equivalent*, but are on the same topic; *on different topics*. However, this definition does not exactly hit the core of our interpretation of similarity because we do not look for

Tweets with a similar meaning (i.e., textual content), but those containing the same emotion, for which the semantic similarity is only an incomplete proxy. Thus, our future research will include finding features that are common for one emotion, while uncommon for the other ones.

Third, a **graph is constructed** consisting of the seeds and unlabelled Tweets as nodes, as well as their respective similarity scores as edges between them. Consequently, the higher the similarity score for two Tweets is, the stronger the edge between them is. Naturally, Tweets that are not found to be similar at all have no edge between them.

Fourth, **graph-based label propagation algorithm** Modified Adsorption (MAD), as presented by Talukdar and Crammer (2009), is applied to the graph. MAD takes a small set of labelled nodes and a large set of unlabelled ones as input along with the similarity scores for all pairs of nodes. From this initial position, the MAD algorithm iteratively propagates the labels along the strongest edges (those edges with highest similarity) to the unlabelled nodes. This results in a probability distribution over all labels for each previously unlabelled node.

We consider MAD best suited for our purpose because it improves the original Label Propagation (Zhu and Ghahramani 2002) and Adsorption (Baluja et al. 2008) algorithms in performance for small sets of seeds and preserves a more variable distribution of labels over the result than Label Propagation. Furthermore, MAD's implementation is easily usable and adaptable as it is publicly available and stores all information like nodes and edge values in text files. Concerning the choice for the family of graph-based semi-supervised learning algorithms, we consider a graph to be the native environment to combine CL and GI because the concept of graphs is well established in both disciplines. Additionally, semi-supervised learning is the only way to receive labelled output from a small set of labelled instances, i.e., the annotation process is less laborious while still guaranteeing high performance of the learning algorithm.

4.3 Data Requirements

In accordance with our research goals, we formulated several requirements for the data sets that can be processed using our proposed method. First, Tweets have to be geolocated. However, this criterion is not sufficient for our purpose because we emphasise the connection between VGI analysis and Urban Planning. Therefore, knowing the location of Tweets is not enough, but our approach also requires that many of them actually concern the location where they were created.

Second, to reduce the amount of data to be annotated, the Tweet set needs to contain as many "emotional" posts as possible. Thus, for testing our approach, we decided to use Tweets in the area of New York City (WGS84 BBOX−74.08, 40.64, −73.89,40.88) during the time of the New York Fashion Week 2014 (6–13 February 2014), and in the area of Boston (WGS84 BBOX−71.21,42.29, −70.95,42.45) during the Boston Marathon event 2013 (15 April 2013). Both of these events attracted comprehensive media attention, including Twitter use, to cause emotional reactions. More, the Tweets were greatly produced in situ (we used geo-referenced Tweets from within the cities, not those talking about the cities), thus representing these particular events well enough. Preliminary results show that we are able to detect and classify emotions in a semi-supervised manner and identify spatio-temporal clusters with the help of CL methods. Detailed results will be presented in a separate follow-up publication.

5 Integration of Emotion Information into Urban Planning Processes

One of the main future tasks for urban and spatial planning is to deal with the impacts and issues of digital spatial data and in the context of big data. Intensified collaboration between the research fields of GIScience and urban planning is essential in the future, particularly in the area of real-world data collected by LBS. At the same time, creative urban scientific approaches have to be developed. The laboratory "urban space" is the test area for developing, adopting and perhaps discarding new planning concepts.

Ideally, the integration in the planning processes takes place during formal consideration, similar to other aspects of "public interest". In an informal planning process, the integration of "urban emotions" can be another channel or aspect in participation processes.

Especially the planning disciplines are in danger of not being able to create adequate new visions or models to react accordingly to influences by these new developments. To enhance "traditional" Planning Support Systems (PSS) like the in the way as it is described by Batty (2014) and Geertman (2002), in which all the known methods and digital techniques are pooled like an early "mash-up" for digital planning with the developments of "wearables", the "quantified self-movement", human and technical sensors for near real-time data gathering opens up a new dimension in urban planning. For instance, the extraction of georeferenced emotions could be used to identify areas where the citizens' well-being is not optimal and where urban planning actions are necessary. The scientific potential of these trends is not exploited in all its varieties and the resulting possibilities for innovative urban analysis and simulation needs to be evaluated in a wide understanding of a "Science of Cities" (Batty 2014).

All these developments urge planners to intensify their scientific efforts in the context of the ecosystem between humans, sensors, the city, and data acquisition, to gather a new quality of urban information and to detect unknown urban patterns. Without understanding these special interactions and relations between all entities, the development of a "smart city" will take place on a technical level and not on an urban planning level. Yet, this does not mean that planners should be increasingly technocratic, but that they are "lawyers" of social aspects in the smart city movement. In addition, the interdisciplinary character of planning has to be one of the main developments of self-conception urban in planning, especially in context of the knowledge society (Streich 2011). Hence, technology aspects constitute a main part of the urban planning's interdisciplinary nature. Only the combination of expertise in technology (which datasets and information can be gathered?), government (how does local administration work?) and sociology/societal skills (which are the societal impacts?) can be a firm basis to develop cities which are really "smart" and improve citizens' lives (Exner 2014). Consequently, future urban planners will need to be far more interdisciplinary in their composition than the equivalent groups producing definitions for the Web only.

6 Discussion and Conclusion

Emotion information of how people perceive their surroundings in the city can build a vital base for innovative urban planning. Thus, *Urban Emotions* provides a human-centred approach for extracting contextual emotional information from technical and human sensor data. The methodology used in our approach comprises four steps: (1) detecting emotions using wristband sensors, (2) "ground-truthing" these measurements using a smartphone-based People as Sensors LBS in near real time, (3) extracting emotion information from crowdsourced data like Twitter, and (4) correlating the measured and extracted emotions. The results are used in the domain of urban planning for decision support and the evaluation of ongoing planning processes.

The uniqueness of Urban Emotions is fourfold: First, the concept improves previous research in that it proposes a **trans-disciplinary approach** combining methods from GIScience, CL and urban sociology by merging the concepts of **semantic, geographic and temporal distance**, and semantic, geographic and temporal clusters. Second, Urban Emotions provides the first application for "**ground-truthing**" emotions in near real time in an urban context using the concept of "People as Sensors". Third, unlike other research efforts, our approach offers direct **feedback to real-world processes** in urban management and planning, and will help to detect previously unseen urban patterns. Finally, the *Urban Emotions* approach is generic so that it is **usable in other areas** like public health, traffic analysis and management, public safety, tourism, etc.

One clear limitation of the methodology for extracting emotion information from VGI is the current reliance on Twitter Tweets, assuming that Tweets are written in situ, i.e., the posts concern the location and time at which they are published. Yet, it has been stated by Hahmann and Burkhardt (2013) that this simplification cannot be assumed to reflect reality.

Apart from this shortcoming, future research includes the optimisation of the People as Sensors user interface, the integration of other base data sources (different sensors and various crowdsourced data repositories), and the creation of a clear set of guidelines for using emotion information in urban planning.

Acknowledgements The authors would like to express their gratitude to the German Research Foundation (DFG—Deutsche Forschungsgemeinschaft) for supporting the project "Urban Emotions", reference number ZE 1,018/1-1 and RE 3,612/1-1. This research has been supported by the Klaus Tschira Stiftung gGmbH.

References

Agirre E, Diab M, Cer D, Gonzalez-Agirre A (2012) Semeval-2012 task 6: a pilot on semantic textual similarity. In: Proceedings of the first joint conference on lexical and computational semantics, Montreal, 7–8 June, pp 385–393

Batty M (2014) Can it happen again? Planning support, lee's requiem and the rise of the smart cities movement. Environ Plan B: Plan Des 41(3):388–391

Baluja S, Seth R, Sivakumar D, Jing Y, Yagnik J, Kumar S, Ravichandran D, Aly M (2008) Video suggestion and discovery for youtube: taking random walks through the view graph. In: Proceedings of the 17th ACM international world wide web conference, Beijing, 21–25 April 2008, pp 895–904

Bollen J, Mao H, Pepe A (2011) Modeling public mood and emotion: twitter sentiment and socio-economic phenomena. In: Proceedings of the fifth international AAAI conference on weblogs and social media (ICWSM), Barcelona, 17–21 July 2011

Bradley MM, Lang PJ (1999) Affective norms for english words (ANEW): instruction manual and affective ratings. Technical Report C-1, The Center for Research in Psychophysiology, University of Florida, pp 1–45

Ekman P, Friesen WV (1971) Constants across cultures in the face and emotion. J Pers Soc Psychol 17(2):124–129

Exner J-P (2014) Smart cities and smart planning. In: Schrenk M et al (eds) Proceedings of RealCORP 2014, Vienna, 21–23 May 2014, pp 603–610

Gartner G (2010) Emotional response to space as an additional concept of supporting wayfinding in ubiquitous cartography. Mapping different geographies. Springer, Berlin, pp 67–73

Geertman S (2002) Participatory planning and GIS: a PSS to bridge the gap. Environ Plan B: Plan Des 29:21–35

Goodchild MF (2007) Citizens as sensors: the world of volunteered geography. GeoJ 69 (4):211–221

Hahmann S, Burghardt D (2013) How much information is geospatially referenced? Networks and cognition. Int J Geogr Inform Sci 27(6):1171–1189

Hauthal, E (2013) Detection, analysis and visualisation of georeferenced emotions. In: Proceedings of the 26th international cartographic conference (ICC 2013), Dresden, Germany, 25–30 Aug 2013

Klettner S, Schmidt M (2012) Visualisierung von Emotionen im Raum. Kartographisches Denken. Springer, Vienna, pp 398–399

Liu B, Zhang L (2012) A survey of opinion mining and sentiment analysis. Mining text data. Springer, USA, pp 415–463

Matei S, Ball-Rokeach SJ, Qiu JL (2001) Fear and misperception of Los Angeles urban space—a spatial-statistical study of communication-shaped mental maps. Commun Res 28(4):429–463

Mohammad SM (2011) Even the abstract have colour: consensus in word–colour associations. In: Proceedings of the 49th annual meeting of the association for computational linguistics, Portland, 19–24 June 2011, pp 368–373

Nold C (2009) Emotional cartography—technologies of the self, ISBN 978-0-9557623-1-4S. 14 May 2014, http://emotionalcartography.net/EmotionalCartography.pdf

Owoputi O, O'Connor B, Dyer C, Gimpel K, Schneider N, Smith NA (2013) In: Proceedings of the conference of the north american chapter of the association for computational linguistics: human language technologies (NAACL HLT 2013), Atlanta, 9–15 June 2013, pp 380–390

Resch B, Britter R, Ratti C (2012) Live urbanism—towards the senseable city and beyond. In: Pardalos P, Rassia S (eds) Sustainable architectural design: impacts on health. Springer, New York, pp 175–184

Resch B (2013) People as sensors and collective sensing—contextual observations complementing geo-sensor network measurements. Progress in location-based services. Springer, Berlin, pp 391–406

Roberts K, Roach MA, Johnson J, Guthrie J, Harabagiu SM (2012) EmpaTweet: annotating and detecting emotions on twitter. In: Proceedings of the international conference on language resources and evaluation (LREC), Istanbul, 21–27 May 2012, pp 3806–3813

Sagl G, Resch B, Hawelka B, Beinat E (2012) From social sensor data to collective human behaviour patterns: analysing and visualising spatio-temporal dynamics in urban environments. In: GI-Forum 2012: geovisualization, society and learning, wichmann Verlag, Berlin, pp 54–63

Salesses P, Schechtner K, Hidalgo CA (2013) The collaborative image of the city: mapping the inequality of urban perception. PLoS ONE 8(7):e68400

Scott B, Neil T (2009) Designing web interfaces: principles and patterns for rich interactions. O'Reilly media, Sebastopol

Streich B (2011) Stadtplanung in der Wissensgesellschaft—Ein Handbuch, 2nd edn. Springer, Berlin

Takhteyev Y, Gruzd A, Wellman B (2012) Geography of twitter networks. Soc Netw 34(1):73–81

Talukdar PP, Crammer K (2009) New regularized algorithms for transductive learning. Machine learning and knowledge discovery in databases. Springer, Berlin, pp 442–457

Tobler WR (1970) A computer movie simulating urban growth in the detroit region. Econ geogr 234–240

Zeile P, Resch B, Exner J-P, Sagl G, Summa A (2014) Urban emotions—Kontextuelle Emotionsinformationen für die Räumliche Planung auf Basis von Echtzeit-Humansensorik und Crowdsourcing-Ansätzen. In: Strobl J, Blaschke T, Griesebner G (eds) Angewandte Geoinformatik 2014. Wichmann Verlag, Heidelberg, pp 664–669

Zeile P, Höffken S, Papastefanou G (2009) Mapping people?—The measurement of physiological data in city areas and the potential benefit for urban planning. In: Proceedings of RealCORP 2009, Vienna, 22–25 April 2009, pp 341–352

Zhu X, Ghahramani Z (2002) Learning from labeled and unlabeled data with label propagation. Technical Report CMU-CALD-02-107, Carnegie Mellon University, Pittsburgh

Citizens as Expert Sensors: One Step Up on the VGI Ladder

Farid Karimipour and Omid Azari

Abstract Volunteered geographic information is one the most significant advances in GIScience following the well-known elements of Web 2.0. It has been mainly looked at as a source of information, and the focus has been on improving its usage, understanding, and quality. This paper opens a discussion on considering VGI as an implicit source of users' experience, which provides general users with solutions to help them take actions like an expert. As a case study, the idea has been applied on estimating the optimum travel time path through information collected by experts. Having introduced the methodology, the results for some examples are presented and discussed.

Keywords Volunteered geographic information · Spatial cognition · Optimum travel time path

1 Introduction

GIScience is experiencing a profound revolution caused by the advances of the digital era and the new pervasiveness of digital spatial data (Sui et al. 2013b). Data acquisition in particular, always a very costly part of geospatial projects, has benefited the most (Goodchild 2009). As an emerging instance, *Volunteered Geographic Information* (*VGI*), coined by Goodchild (2007), conceptualizes the use of unbeatable power, knowledge, expertise and ubiquity of general public in spatial data acquisition. VGI refers to "a range of geo-collaboration projects in which individuals voluntarily collect, maintain and visualize information" (Thatcher 2013). Among practical implementations are Open Street Map (OSM) and Wikimapia, whose data are provided by their users.

F. Karimipour (✉) · O. Azari
Department of Surveying and Geomatics Engineering, College of Engineering,
University of Tehran, Tehran, Iran
e-mail: fkarimipr@ut.ac.ir

G. Gartner and H. Huang (eds.), *Progress in Location-Based Services 2014*,
Lecture Notes in Geoinformation and Cartography,
DOI 10.1007/978-3-319-11879-6_15

Most VGI-related studies have considered VGI as a source of information, and have focused on improving its usage, understanding, or quality (Goodchild and Li 2012; Karimipour et al. 2013; Roche et al. 2012; Sui et al. 2013a). But VGI is also an implicit source of its users' experience: mining the information collected by a group of users who are experts in a specific spatial action may provide better solutions for other users, allowing them to take that same action the way an expert would. This viewpoint has not been well considered in pertinent literature. An exception is the effort to move VGI from being mere information to becoming a service provider that supports actions (Savelyev et al. 2011; Thatcher 2013). These so-called Volunteered Geographic Services (VGS) use volunteered geographic information to aggregate, correlate, and present information in a manner useful for specific services (e.g. rescue services, location-based services, etc.). There are also research directions to correlate spatial data obtained from the general public and human social behavior, and extract large-scale patterns like human physical travel, communication travel, and environmental structure from individual and urban-oriented perspectives. (Andrienko et al. 2013; Naboulsi et al. 2013; Paraskevopoulos et al. 2013; Yuan and Raubal 2013; Yuan et al. 2012). In both these cases, however, the users are not necessarily experts (especially in the second case), and thus the information provided by them may not be fully reliable. Furthermore, in the case of VGS the users are aware of the specific pre-defined action, and consciously collect information only for that purpose; whereas in the latter the information is extracted from the daily activities of people, and could thus more accurately be called crowdsourcing rather than VGI, as the information is not voluntarily collected.

In this paper, we introduce the idea of extracting the experience that exists in the volunteered geographic information provided by experts in a specific action. This is an effort made to use VGI towards the belief that "GI science deals with the formal modelling of spatial process and interaction of humans with the environment in space and time" (Frank 2000). As a case study, we represent the initial results of ongoing research on estimating the optimum travel time path through information collected by experts.

The rest of the paper is structured as follows: Sect. 2 introduces the research idea in more detail. In Sect. 3, we show how the proposed idea may be applied on finding the optimum travel time path, as a frequently used case study, and present and evaluate the results for some examples. Finally, Sect. 4 contains the concluding remarks and directions for future work.

2 VGI, Spatial Cognition and Experience

Flanagin and Metzger (2008) believe that the data acquired by specialists is not necessarily always accurate. Though unreliable, non-specialists sometimes provide more accurate data, which is the idea behind VGI. However, a major challenge of VGI is its quality and verification, for which no generally accepted definitions have been proposed. VGI is an integrated source of spatial information collected by a general public with different levels of knowledge, education, and abilities in spatial data acquisition. Furthermore, VGI is more based on human cognition than on measurement, as it is collected by ordinary people with only a vague view of space and geography. It is therefore more reasonable to evaluate its quality based on a *degree of truth* rather than accuracy (Longueville et al. 2012). Flanagin and Metzger (2008) proposed the concept of *credibility* for verifying VGI, which is a conceptual variable defined based on *believability*, *trustworthiness* and *expertise* (Hovland 1953), and is measured in relation to the individuals. This in turn means that the users' experience has a direct effect on the reliability of the collected information.

On the other hand, volunteered geographic information is collected by people who live in the environment and interact with it through their spatial cognition (Lynch 1960). They know the streets, buildings, parks, etc., and they also cognitively know how to perform certain spatial interactions with the environment in an effective manner. In other words, living in an environment results in a cognitive map that not only contains spatial elements, but implicitly contains experience on how to perform actions, which are obviously different from person to person (Raubal 2008): The longer you live in an environment, the better a path you would choose to get from an origin to a destination, as you know more primary and secondary roads. One of the aspects of *crowd quality* (*CQ*) introduced by Exel et al. (2012) to describe the quality of VGI is the user's quality, which is defined based on his/her local knowledge (i.e. knowledge about the environment) and experience. It is in close relation to the concept of *affordance* that describes how people visually perceive their environment (Gibson 1979).

The above discussion lead to the proposed idea of the paper, which can be summarized as follow: Interactions between humans and their environment lead to experience, which consequently results in more reliable information provided. In other words, the information obtained from people with more experience is more reliable as a source of information and experience.

Goodchild (2007) introduced three types of sensors (Fig. 1):

- Static sensors that capture certain measurements from their local environments.
- Sensors carried by humans, vehicles, animals, etc. in order to understand the variations of a certain environmental factor, or its effect on specific phenomena.
- Humans themselves, equipped with five senses and with the intelligence to compile and interpret what they sense. VGI is an effective use of the network of this type of sensor, with over 6 billion components.

The proposed idea of the paper can be considered a forth type of sensor, an integration of the 2nd and the 3rd: sensors that are carried by people, with the intelligence, experience and five senses, which affect the collected information and its quality and reliability. We propose extracting the experience that exists within the spatial information provided by people who are experts in a specific action, and later deploy this experience to provide information to other users performing the same action.

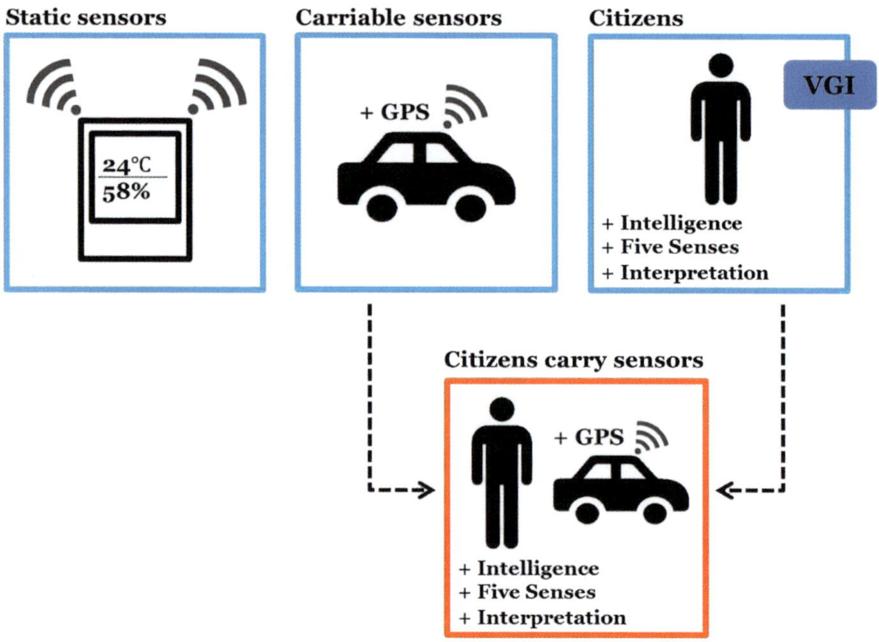

Fig. 1 *Top* Three types of sensors introduced by Goodchild (2007). *Bottom* Integration of the 2nd and the 3rd types of sensors, which results in expert sensors

3 Case Study: The Optimum Travel Time Path

This section, as a case study, shows how the data collected by expert drivers can provide better optimum travel time paths.

Optimum path finding is a classical problem for which several algorithms have been proposed (Bellman 1958; Dijkstra 1959; Hart et al. 1968). Among different concepts of "optimality" (e.g. length, cost, time, etc.), which have resulted in different solutions (Frank 1969; Nie and Wu 2009; Shirabe 2008), the shortest travel time is a practical, exhaustively studied variant (Fu and Rilett 1998; Hall 1986; Jigang et al. 2011; Miller-Hooks 1998; Orda and Rom 1990; Pattanamekar et al. 2003; Sung et al. 2000). In this case, the travel time on each edge of the network may be estimated through: (1) a fixed average travel time assigned to that edge; (2) a time-dependent travel time lookup table assigned to the edge, which has been extracted from historical traffic data; and (3) online traffic data transmitted to the optimum path calculation device. Although they are quite helpful to an extent, a level of uncertainty exists in the results, which leads to reduced reliability. Especially if the users are knowledgeable about the environment and its traffic patterns, they may not fully trust the results of the fastest paths calculated merely through stochastic algorithms. Furthermore, it is obvious that in the first and second cases

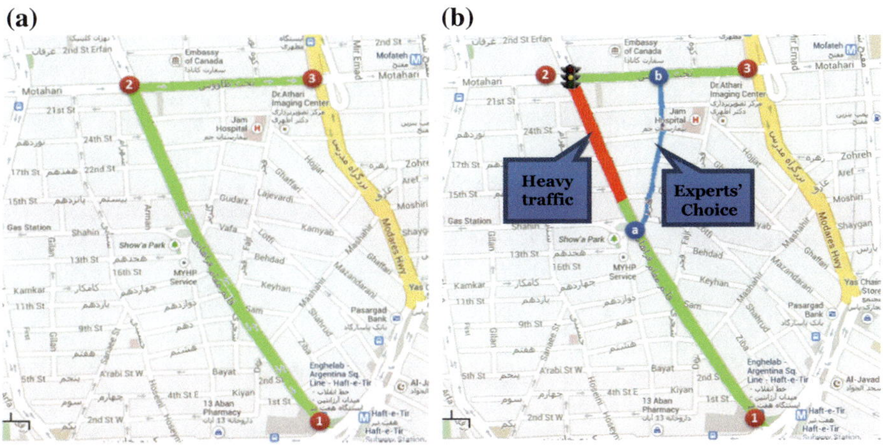

Fig. 2 The optimum travel time path between nodes #1 and #3: **a** Normal traffic flow. **b** The deviation taken by experts because of the heavy traffic

the travel time has enough uncertainty to not be close enough to the current situation. Even in the third case, the traffic situation on an edge may drastically change by the time the user reaches there.

In this section, we represent the initial results of ongoing research on estimating the optimum travel time path through information collected by experts, who have used their experience and spatial cognition to choose appropriate paths (Fig. 2), grouped by days of the week and times of day. A model is introduced to consider the temporal aspects of the problem, which is then deployed to use the data provided by experts in order to find an optimum travel time path that would be taken by someone who knows the area and its traffic patterns. The implementation results for some examples are presented and discussed.

3.1 Fu's Model for Time-Dependent Path Finding

In time-dependent path finding, each edge is equipped with time-dependent travel time information. In Fig. 3, if one reaches the point i at the time instant t, it takes $g_{ij}(t)$ to pass the edge ij and arrive at the point j at $t + g_{ij}(t)$. The shortest path from i to N is obtained by finding the minimum of the following function (Bellman 1958; Cooke and Halsey 1966):

$$f_i(t) = \min\big(g_{ij}(t) + f_j(t + g_{ij}(t))\big), \quad i = 1, 2, \ldots, N$$

$$j \neq i, f_N(t) = 0$$

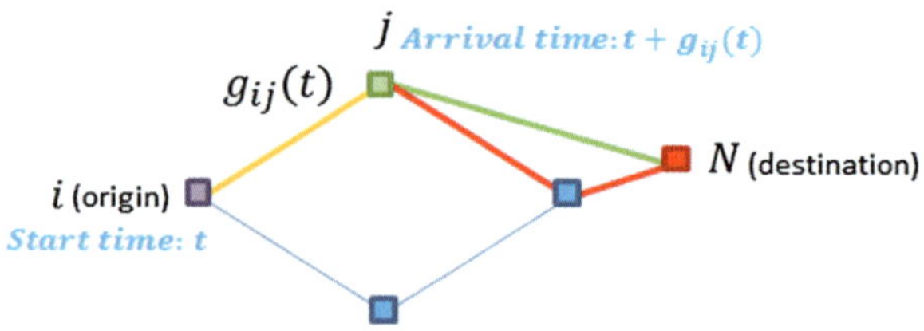

Fig. 3 Modeling the optimum travel time path

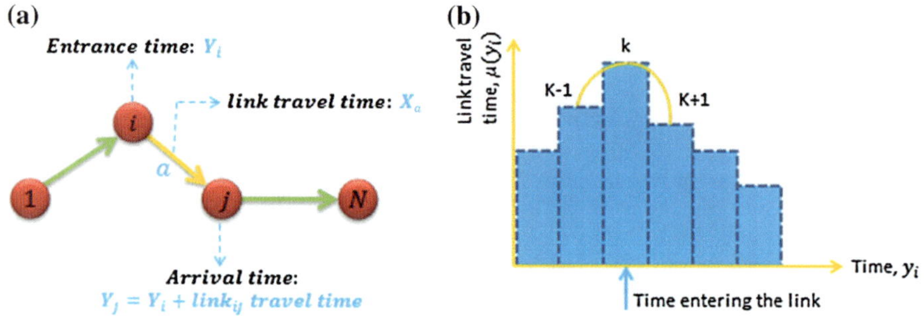

Fig. 4 The model proposed by Fu and Rilett: **a** The elements of the model;**b** Estimation of E[Y_i]

In absence of online traffic data, if the edges' weights (i.e. travel times) are only computed through averaging of historical data, the stochastic characteristics of the problem are completely ignored. Fu and Rilett (1998) proposed a model that considers both dynamic and stochastic issues on traffic networks. Assume travelling from point *1* to *N* in the example illustrated in Fig. 4a. For each edge, e.g. the edge *a* between the points *i* and *j*, the entrance and travel time are stochastic values oscillating around the average, as the model is both dynamic and stochastic; and:

- Y_i is the entrance time to edge *a*, between points *i* and *j*.
- X_a is the travel time on edge *a*, which is a function of Y_i.
- Y_j is the exit time of edge *a*, or the entrance time to point *j*.

Obviously $Y_j = Y_i + X_a$ As the time values are stochastic, they are replaced with their mathematical expectations:

$$E[Y_j] = E[Y_i] + E[\mu_{X_a}(Y_i)]$$

To estimate the expected value of $[Y_i]$, the continuous travel time on edge *a* is discretized (Fig. 4b), and a second-order polynomial is fitted to the data stored for the entrance, as well as its previous and next time intervals. The value of $E[\mu_{X_a}(Y_i)]$ is estimated through a second-order Taylor expansion at $t = E[Y_i]$:

$$E\left[\mu_{X_a}(Y_i)\right] = \mu_{X_a}(E[Y_i]) + \frac{1}{2}\mu''_{X_a}(E[Y_i]) \times Var[(Y_i)]$$

If the first-order Taylor expansion is used, i.e. $E\left[\mu_{X_a}(Y_i)\right] = \mu_{X_a}(E[Y_i])$, the stochastic values are replaced with the averages of the travel times; thus the problem becomes dynamic, but not stochastic. In case of second-order expansion, however, the model remains stochastic, as the variance is involved (Fu and Rilett 1998). The average and variance values can be computed using the historical data provided by GPS at discrete time instants.

3.2 Involving the Experts

As discussed, experts rely more on their own knowledge and experience when finding the optimum path, instead of on algorithmic methods. Here, we assume the quality of the optimum path is improved by deploying the data collected by the expert drivers, who knows the time-dependent optimum path. Thus, their data is considered to be more reliable, creditable and trustworthy. In order to extract this knowledge, we propose storing the paths taken by experts for their travels on the network for different days of the week and different times of day, and using this information to assign weights to the network's edges. For this, we discretized the continuous time to 7 days and classified each day into several time intervals based on the traffic pattern. Then, the travel time of each edge for each time class was estimated through averaging the travel time of the experts who passed that edge in that time class. Applying Fu's model on this network will provide the optimum path proposed by the experts.

3.3 Results and Discussion

The proposed idea has been implemented for a sample dataset. We used data provided by taxi drivers as the experts who cognitively know the time-dependent optimum path. Figure 5 compares the results of three paths proposed by Fu's model based on the historical and the experts' data for certain time instants. The deviations of the experts are highlighted. In the third example, note how the experts have avoided the traffic light!

We do not claim that this is the exact optimum path, as it is merely tentatively proposed by users whose spatial knowledge about the environment may very well not be complete. There could also be, say, a car accident that has happened on the

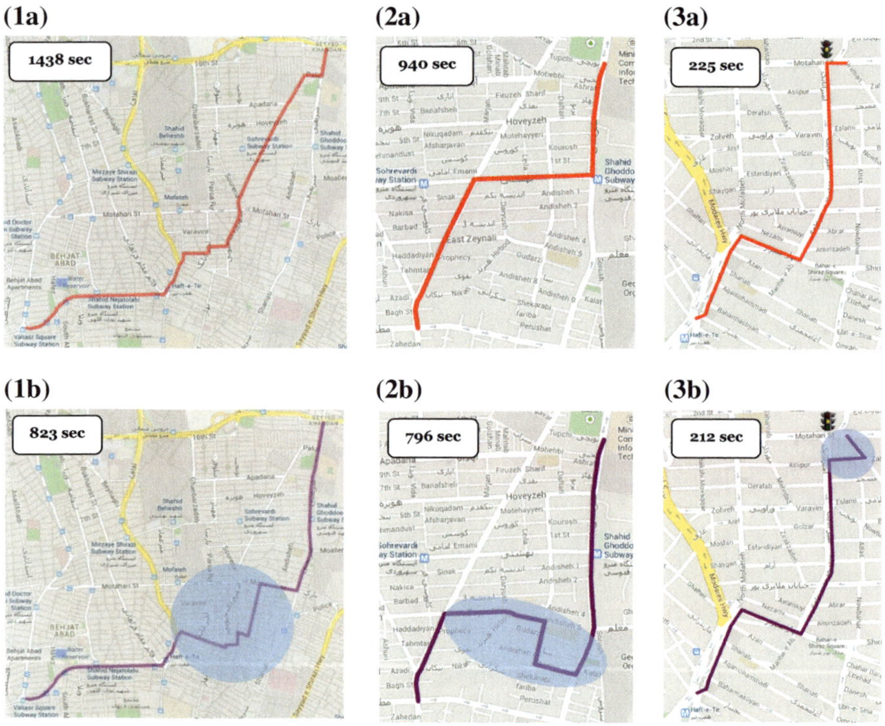

Fig. 5 The Fu's optimum travel time path using historical data (*top*) and the information obtained from experts (*bottom*)

path proposed by the experts, which could have been identified by online traffic information. Nevertheless, we are relying on people who are frequently travelling in the area, instead of relying purely on computational algorithms.

4 Conclusion

In this paper we opened a discussion on considering VGI as an implicit source of users' experience, which provides general users with solutions to help them take actions like an expert. The initial implementation results for a case study, i.e. estimating the optimum travel time path through information collected by experts, illustrates the elegancy of the approach. However, there are still several issues to be considered: in the present case we already knew that the information used had been collected by experts. In the absence of such knowledge, i.e. information provided by a general public, a degree of truth and credibility must be assigned to the inputs (Flanagin and Metzger 2008; Hovland 1953). In our case study, we can provide such indices by measuring the correlation and similarity of the paths (Yuan and

Raubal 2013; Yuan et al. 2012). Here, we use the "travel time" as the "optimality" parameter, so the taxi drivers were considered as experts. However, for other "optimality" parameters (e.g. security, safety, noise, pollution, etc.) other users may be considered as experts. Ultimately, an integration of expert information with computational methods and auxiliary data may provide more accurate and reliable solutions. In the optimum path example, integration of the experts' data with online traffic data would allow a user to act as an expert who is travelling on a network based on their experience, while also being informed about unexpected traffic situations and thus able to make changes to the path.

References

Bellman R (1958) On a routing problem. Q Appl Math 16:87–90

Cooke KL, Halsey E (1966) The shortest route through a network with time-dependent internodal transit times. J Math Anal Appl 14:493–498

Dijkstra EW (1959) A note on two problems in connexion with graphs. Numer Math 1:269–271

Exel MV, Dias E, Fruijtier S (2012) The impact of crowd sourcing on spatial data quality indicators. In: Proceedings of the GIScience. Zurich, Switzerland

Flanagin AJ, Metzger MJ (2008) The credibility of volunteered geographic information. GeoJournal 72:137–148

Frank AU (2000) Geographic information science: new method and technology. J Geogr Syst 2:99–105

Frank H (1969) Shortest paths in probabilistic graphs. Oper Res 17:583–599

Fu L, Rilett LR (1998) Expected shortest paths in dynamic and stochastic traffic networks. Transp Res Part B: Methodological 32:499–516

Andrienko G, Andrienko N, Bak P et al (2013) Visual analytics of movement. Springer, Heidelberg

Gibson J (1979) The ecological approach to visual perception. Houghton Mifflin Company, Boston

Goodchild MF (2007) Citizens as sensors: the world of volunteered geography. GeoJournal 69:211–221

Goodchild MF (2009) Geographic information systems and science: today and tomorrow. Annals GIS 15:3–9

Goodchild MF, Li L (2012) Assuring the quality of volunteered geographic information. Spat Stat 1:110–120

Hall RW (1986) The fastest path through a network with random time-dependent travel times. Transp Sci 20:182–188

Hart PE, Nilsson NJ, Raphael B (1968) A formal basis for the heuristic determination of minimum cost paths. IEEE Trans Syst Sci Cybern 4:100–107

Hovland CI, Janis IL, Kelley JJ (1953) Communication and persuasion. Yale University Press, Connecticut

Jigang W, Jin S, Ji H, Srikanthan T (2011) Algorithm for time-dependent shortest safe path on transportation networks. Procedia Comput Sci 4:958–966

Karimipour F, Esmaeili R, Navratil G (2013) Cartographic representation of spatial data quality parameters in volunteered geographic information. In: The 26th international cartographic conference (ICC)

Longueville BD, Ostländer N, Keskitalo C (2012) Addressing vagueness in volunteered geographic information (Vgi)—a case study. Int J Spat Data Infrastruct Res 5

Lynch K (1960) The image of the city. MIT Press, Cambridge

Miller-Hooks ED (1998) Least possible time paths in stochastic, time-varying networks. Comput Oper Res 25:1107–1125
Naboulsi D, Fiore M, Stanica R (2013) Human mobility flows in the city of abidjan. 3rd International conference on the analysis of mobile phone datasets (NetMob2013), pp 1–8
Nie Y, Wu X (2009) Shortest path problem considering on-time arrival probability. Transp Res Part B: Methodological 43:597–613
Orda A, Rom R (1990) Shortest-path and minimum-delay algorithms in networks with time-dependent edge-length. J ACM 37:607–625
Paraskevopoulos P, Dinh T, Dashdorj Z et al (2013) Identification and characterization of human behavior patterns from mobile phone data. 3rd international conference on the analysis of mobile phone datasets (NetMob2013)
Pattanamekar P, Park D, Rilett LR, Lee J, Lee C (2003) Dynamic and stochastic shortest path in transportation networks with two components of travel time uncertainty. Transp Res Part C: Emerg Technol 11:331–354
Raubal M (2008) Wayfinding: Affordances and Agent Simulation. In: Shekhar S, Xiong H (eds) Encyclopedia of gis. Springer, US, pp 1243–1246
Roche S, Mericskay B, Batita W, Bach M, Rondeau M (2012) Wikigis basic concepts: Web 2.0 for geospatial collaboration. Future Internet 4:265–284
Savelyev A, Xu S, Janowicz K, Mülligann C et al (2011) Volunteered geographic services: developing a linked data driven location-based service. Proceedings of the 1st ACM SIGSPATIAL international workshop on spatial semantics and ontologies, New York, USA
Shirabe T (2008) Minimum work paths in elevated networks. Networks 52:88–97
Sui D, Elwood S, Goodchild M (eds) (2013a) Crowdsourcing geographic knowledge-volunteered geographic information (vgi) in theory and practice. Springer, Heidelberg
Sui D, Goodchild M, Elwood S (2013b) Volunteered geographic information, the exaflood, and the growing digital divide. In: Sui D, Elwood S, Goodchild M (eds) Crowdsourcing geographic knowledge-volunteered geographic information (vgi) in theory and practice. Springer, Heidelberg, pp 1–12
Sung K, Bell MGH, Seong M, Park S (2000) Shortest paths in a network with time-dependent flow speeds. Eur J Oper Res 121:32–39
Thatcher J (2013) From volunteered geographic information to volunteered geographic services. In: Sui D, Elwood S, Goodchild M (eds) Crowdsourcing geographic knowledge. Springer, Netherlands, pp 161–173
Yuan Y, Raubal M (2013) Investigating the distribution of human activity space from mobile phone usage. mobile Ghent 2013, Belgium
Yuan Y, Raubal M, Liu Y (2012) Correlating mobile phone usage and travel behavior—a case study of harbin, china. Comput Environ Urban Syst 36:118–130

LBS-Based Dilemma Zone Warning System at Signalized Intersection

Yi Li, Junhua Wang and Lanfang Zhang

Abstract In the field of traffic engineering, drivers' confusion during yellow interval is a great danger to traffic safety. To help drivers solve such problem, an LBS-based dilemma zone warning system at signalized intersection is designed. Different from other dilemma zone solving systems, this is an active safety system, which makes it possible to help drivers take active actions according to reliable warnings. An algorithm based on data from Differential Global Positioning System (DGPS) and traffic signal detector (TSD) was built in the single-chip microcomputer of Dedicated Short Range Communication (DSRC). Corresponding voice warnings ("stop" or "go") would be sent to drivers by on-board voice alerter. Besides, communication delay and drivers' reaction time were taken into consideration in the design of warning areas ahead of dilemma zone. To reveal characteristics of reaction time, a field test has been conducted and the reaction time is shorter than current standard. When vehicles are detected in different warning areas, drivers would get alarmed at the beginning of yellow interval. Moreover, this system has been verified feasible in theory by a case study at signalized intersection.

Keywords LBS · Dilemma zone · DSRC · Warning system · Drivers' reaction time

1 Introduction

Traffic safety is now a worldwide topic. Although the mileage of road intersections is small, but the number of accidents happened at intersections is greater than that of sections. In 2010, 219,521 accidents happened in China, among which 39,945 accidents occurred near intersections. In America, the amount of intersection accidents was about 36 % of the total number of accidents. In Japan, this rate was

Y. Li · J. Wang (✉) · L. Zhang
School of Transportation Engineering, Tongji University, 201804 Shanghai, China
e-mail: benwjh@163.com

G. Gartner and H. Huang (eds.), *Progress in Location-Based Services 2014*,
Lecture Notes in Geoinformation and Cartography,
DOI 10.1007/978-3-319-11879-6_16

42.2 %. This shows that intersection is a high-risk spot that affects drivers' behavior and decisions. Papaioannou (2007) studied driving behavior at signalized intersections in Greece. The findings of this research indicated that a large percentage of drivers during yellow interval were caught in a dilemma zone due to high approaching speed and aggressive behavior. Gates et al. (2007) evaluated the behavior of vehicles between 2.0 and 5.5 s upstream of signalized intersections at the start of yellow interval. Installed cameras on roadside helped to analyze the deceleration rate and brake-response times of first-to-stop vehicles. It indicated that deceleration increased with approaching speed and drivers' sensation time. Besides, some challenging circumstances made drivers choose to stop in yellow interval. This means that unfamiliar with light interval of a signalized intersection leads drivers to make wrong decisions, like speeding, aggressive driving or slam braking. All of these will deteriorate upstream traffic environment of signalized intersections, especially in dilemma zone.

Dilemma zone is an area that many drivers are familiar with. In this region, drivers have to make a choice: to stop or to keep going in yellow interval. The phenomenon of dilemma zone near signalized intersection has been studied by many researchers (Urbanikand and Koonce 2007). Traditional method was to minimize the number of vehicles trapped in dilemma zone (Zegeer 1978; Pant and Cheng 2001).

To improve former methods and minimize the influence of dilemma zone, several approaches have been proposed (Li et al. 2009; Lei 2010), such as minimizing enforcement tolerance for photographing red light running and adjustment of phase change interval of yellow and red time. However, the extension of green light is limited, so that Sharma et al. (2007) used an economic evaluation approach to handle the balance of safety and delay, which took traffic conflicts and induced delay cost into consideration. Moreover, a dilemma zone hazard function estimating procedure was developed to obtain the probability of traffic conflict occurring (Sharma et al. 2011). Based on lane-by-lane and vehicle length, Middleton et al. (2011) evaluated a detection-control system (D-CS) at eight sites, which obtained information from detectors that were located upstream of intersections to extend green interval and to monitor individual vehicles on intersection approach. It had high success rate in preventing crashes and showed salient feature on trucks. Tong et al. (2009) introduced a concept of "generalized yellow light dilemma zone" (GDZ), by which all the violation vehicles could be judged out. With this new concept, dilemma zone could be included in intersection collision avoidance system. Heng and Zhixia (2009) conducted a proof-of-concept development of an innovative methodology for modeling the dynamic dilemma zones using video-capture techniques and time-based trajectory data. The lookup chart provided a tool for identifying real-time location and length of dilemma zone, but more signalized intersections should be analyzed to ensure the extensive application of this method.

With the development of advanced roadside communication devices, short-range communication units have been applied in car-vehicle communication system to improve its real-time performance and traffic environment. McCoy and Pesti (2003) compared two methods of dilemma-zone protection: a conventional design with advance detectors at multiple locations and a new design with advance detection at only one location and advance warning flashers on signs. The limitation of the new design in low volumes was solved by a modified model. An in-vehicle dilemma zone warning system was developed by Moon et al. (2003). Field tests were also conducted at signalized intersections, which verified the good performance of such system in reducing red light violations and collisions. With large data collection range of DSRC-only system, Cho et al. (2012) established an advanced driving-assistance system. Besides, CO_2 emissions of per vehicle would also be reduced by nearly 7 % in the case of v/c = 0.7 (Li et al. 2009) with the help of such system.

The purpose of this research is to establish a decision support system to suggest drivers what to do when they face with yellow light at signalized intersections. Different from former systems, this dilemma zone warning system is a driver-targeted system. With the help of the communication between on-board devices and roadside devices, drivers don't need to "hurry up" or "stop immediately" when they are approaching signalized intersections. They can take actions according to system warnings. This would not only improve driving safety, but also reduce fuel consumption. Formulas and critical value are presented in this paper to form the flowchart of this system.

2 Definition of Yellow Interval Dilemma Zone

Yellow interval dilemma zone is a hazardous area for drivers. Driving towards a signalized intersection, drivers would take different actions in red, yellow and green light intervals. Undoubted decision (stop or go) can be made by drivers during red and green intervals. But when yellow light begins, drivers' decision will be influenced by both the vehicle speed and the distance away from the intersection. In yellow interval, if the distance away from stopping line is shorter than threshold value (D_s), then vehicle couldn't stop safely before the light turns to red. On the other hand, if vehicles couldn't pass the clearing line in yellow interval, drivers would accelerate or be trapped in the intersection at the end of yellow interval. Both situations are dangerous to drivers and traffic flow. Dilemma zone is shown in Fig. 1.

In Fig. 1, vehicle is assumed to be x (meters) away from intersection. According to vehicle kinematics theories, dilemma zone can be obtained from the two following circumstances.

Circumstance 1 x is long enough for drivers to stop car before stopping line, then they would choose to decelerate when traffic light turns yellow.

$$x \geq D_s = \frac{V_0}{3.6} t_0 + \frac{V_0^2}{25.92a} (\mathrm{m}) \tag{1}$$

Circumstance 2 x is too short for drivers to stop car before stopping line, then they would choose to pass the intersection in yellow interval.

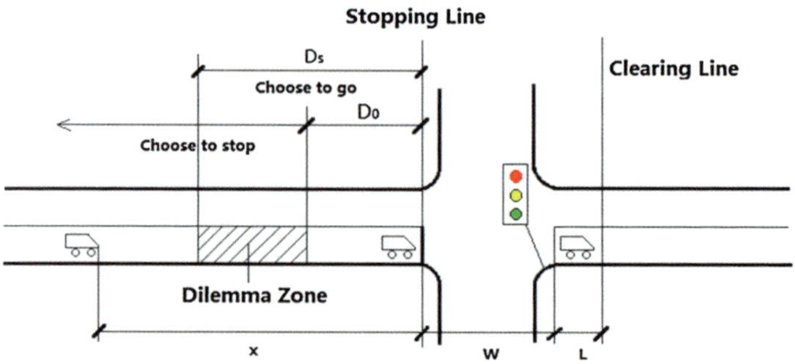

Fig. 1 Definition of dilemma zone

$$x + W + L \leq D_0 + W + L = \frac{V_0}{3.6}(t_0 + \theta)(\mathrm{m}) \quad (2)$$

$$x \leq D_0 = \frac{V_0}{3.6}(t_0 + \theta) - W - L(\mathrm{m}) \quad (3)$$

where D_s is stopping sight distance that depends on both drivers' reaction characteristics and vehicle braking distance. D_0 is the critical distance if drivers want to get to clearing line safely at the end of yellow interval. t_0 is drivers' reaction time. Other parameters are explained as follows,

x—vehicle's initial distance away from stopping line at the beginning of yellow interval (m);
W—width of intersection (m);
L—vehicle length (6 m, according to Chinese Highway Technical Standard);
v_0—initial speed of tested vehicle at the beginning of yellow interval (km/h);
a—initial speed of tested vehicle at the beginning of yellow interval (km/h);
θ—the length of yellow interval.

If the vehicle happens to be in the overlapping region, then the driver could not stop safely or pass the intersection timely. Therefore dilemma zone can be described as $D_0 \leq x \leq D_s$. The foundation of this inequality is

$$D_0 < D_s \quad (4)$$

$$\frac{V_0}{3.6}(t_0 + \theta) - W - L < \frac{V_0}{3.6}t_0 + \frac{V_0^2}{25.92a} \quad (5)$$

$$\theta < \theta_0 = \frac{V_0}{7.2a} + \frac{3.6}{V_0}(W + L) \quad (6)$$

If yellow light interval (θ) is larger than θ_0, there would be no dilemma zone. Drivers can either stop safely before stopping line or pass intersection at a constant speed, regardless of special cases.

3 LBS-Based Dilemma Zone Warning System (DZWS)

3.1 Devices and Layout

Due to drivers' limitation of sensing the changing moment traffic light, assistant measures are needed to help drivers to make a choice: to stop or to go. LBS devices [Roadside Unit (RSU) and On-board Unit (OBU)] detect yellow interval dilemma zone based on traffic signal detector (TSD), roadside device (DSRC) and onboard device (DGPS and alerter).

The system layout of all devices is shown in Fig. 2.

TSD is a built-in device that can detect the start and end of yellow light. This is now a common technology used in ITS (Intelligent Transportation System). However without OBU and roadside communication equipments, current devices can only tell drivers the rest time of red or green interval with countdown signals (Fig. 3). Because yellow interval is short (3–4 s or shorter), no countdown signal is set on yellow light.

Some improved equipment can extend yellow interval with roadside devices and specific algorithm. However, drivers cannot get an active traffic warning based on their own vehicle speed, own position and signal interval. To overcome this, OBU (DGPS and voice alerter) and roadside communicator (DSRC) are included in DZWS. The positional accuracy and update frequency of several positioning technologies are listed in Table 1.

Among these positioning technologies, GPS is currently widely used on smart phones and on-board navigation systems, but its accuracy is not enough for signalized intersection warning system. DGPS is an enhanced technology, which can decrease location error with the help of base station. This technology is now available in China. Inertial navigation technology predicts vehicle location with gyroscope and other vehicle state sensors. This solution highly relies on accurate

Fig. 2 System layout of LBS devices

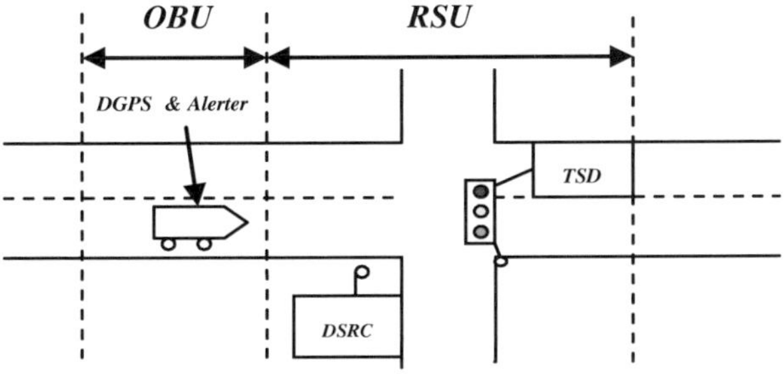

Fig. 3 Countdown traffic signals

Table 1 Parameters of different positioning technologies

Positioning technology	Positional accuracy (m)	Update frequency (Hz)
GPS	<5	1–10
DGPS	<1	1–10
Inertial navigation	<1	100–200/<35/<50

background algorithm and high frequency transmission devices, which can only be available in some professional research or tests. Therefore DGPS is chosen in this LBS-based dilemma zone warning system and it can be widely used in the near future.

The other important device in this system is DSRC. This device can automatically communicate with moving vehicle or/and traffic signal controller (one-way or two-way communication). It can receive and send signals from short-range to medium-range wirelessly. Some parameters of DSRC are shown in Table 2.

With these two main devices and common traffic signal detector, drivers can get alarmed of yellow light and decision aid by an on-board alerter. Figure 4 shows the diagram of the system.

Table 2 Parameters of DSRC

Attribute	Description
Operating mode	WAVE (1,609.3, 1,609.4, 802.11p, P1609.2 with optional libraries)
Frequencies	5.725–5.850 GHz
Data rates	3–27 Mbps (10 MHz channels), 6–54 Mbps (20 MHz channels)
Tx output power	10–20 dBm (rate-dependent), in 1 dB steps measured at antenna connector

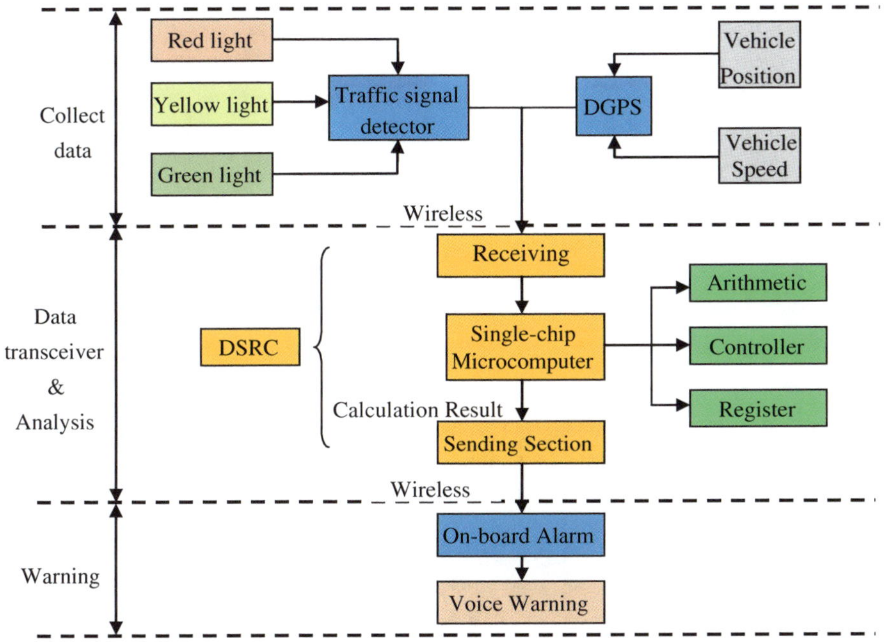

Fig. 4 LBS-based dilemma zone warning system diagram

3.2 Time Delay of DZWS

3.2.1 Communication Delay

In ideal situation, drivers can get out of dilemma zone at signalized intersection with the assistant of LBS device system. However, due to the delay of communication system, drivers cannot immediately get alarmed at the beginning of yellow light.

Communication delay of this system consists of four parts: time delay in data collection, time delay in data receiving, time delay in data analysis and sending delay, which are listed in Table 3.

3.2.2 Drivers' Reaction Time

In former research, drivers' reaction time to stimulation is divided into two parts: sensation time and reaction time, which were set to 1.5 and 1.0 s respectively in

Table 3 Time delay of each progress

	DGPS (s)	DGPS ↓ DSRC (s)	DSRC (model calculation) (s)	Trafficsignaldetector ↓ DSRC (s)	DSRC ↓ Alerter (s)	Total delay (s)
Data collection	1 s	–	–	–	–	1
Data receiving progress	–	0.05 s	–	0.05 s	–	0.1
Data analysis	–	–	0.05 s	–	–	0.05
Data result sending	–	–	–	–	0.05 s	0.05
Total						1.2

Chinese Highway Standard. However, the total time of these two parts would change with different speed, so that a modified reaction time is introduced to level up the accuracy of the warning system.

30 drivers (20 male and 10 female), who all have more than 3 years driving experience, were chosen to test their reaction time to voice stimulation. In the experiment, they were asked to brake when they got a voice alarm. The tests were taken in three circumstances: road design speed of 60, 80 and 100 km/h. Tests results are presented in Figs. 5, 6, and 7.

According to the distribution of drivers' reaction time to voice alarm in different speed, drivers' reaction time climbs with the increase of speed. The 85th quantile of drivers' reaction time in 60, 80 and 100 km/h are 1004, 1084 and 1120 ms respectively. Better than current standard in China, this result presents an obvious

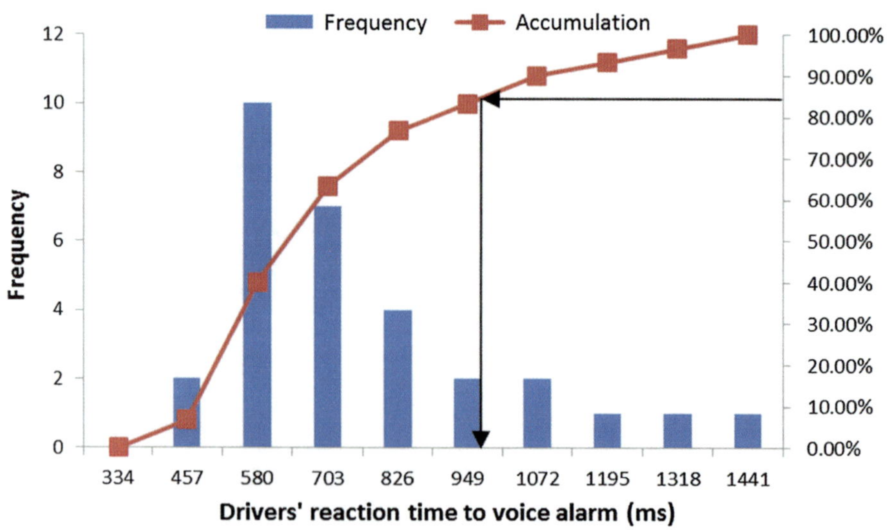

Fig. 5 Drivers' reaction time to voice alarm (60 km/h)

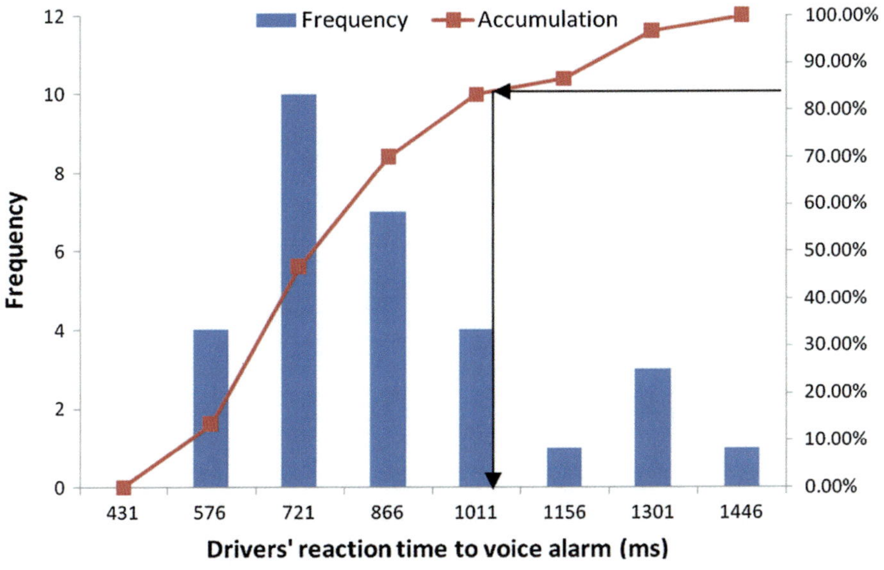

Fig. 6 Drivers' reaction time to voice alarm (80 km/h)

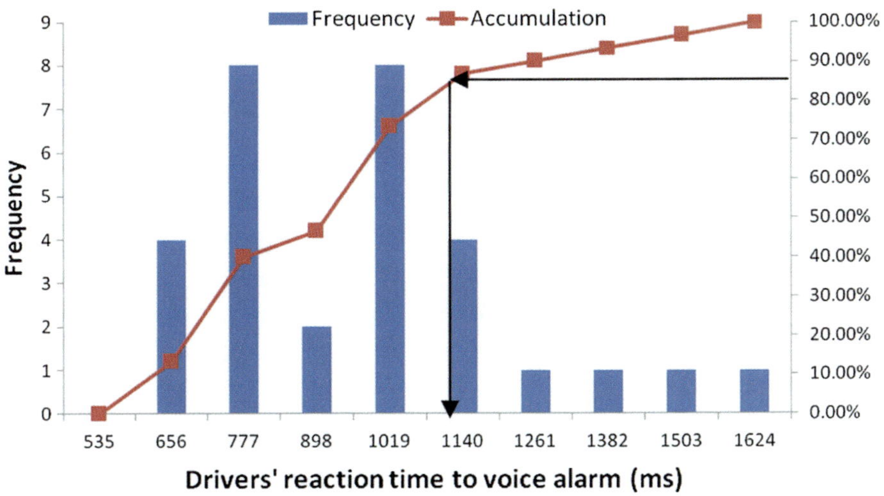

Fig. 7 Drivers' reaction time to voice alarm (100 km/h)

change of reaction time with speed and the value is much smaller than 2.5 s (Chinese standard). Based on this field test, the warning system would be more accurate and targeted to drivers.

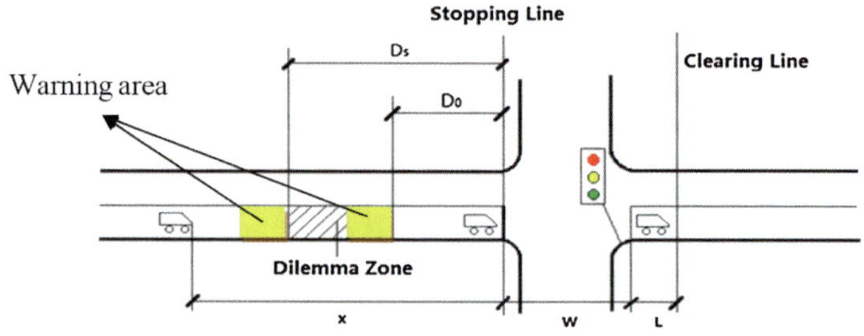

Fig. 8 Warning areas ahead of dilemma zone

3.3 Warning Area

Owing to the time delay of communication and drivers' reaction time, the warning moment should be T = 1.2 s + 1 s = 2.2 s earlier than yellow interval. Corresponding warning areas are set ahead of dilemma zone (see Fig. 8).

The time delay of communication and drivers' reaction time are independent of each other and the sum of them decides the length of warning area. D_{delay} is assumed as vehicle's travel range during T.

$$D_{delay} = T \times \frac{V_0}{3.6} = 0.28V_0T(\mathrm{m}) \tag{7}$$

With this value and other parameters, some critical values can be got as follows,

$$D_s = 0.0129V_0^2 + 0.28V_0T(\mathrm{m}) \tag{8}$$

$$D_0 = 0.28V_0T + 0.28V_0\theta - W - 6(\mathrm{m}) \tag{9}$$

$$\theta_0 = 0.046V_0 + \frac{3.6}{V_0}(W + 6)(\mathrm{s}) \tag{10}$$

3.4 Case Study

In this study, vehicles' position is obtained by DGPS. The location data and traffic light interval of the selected intersection are stored in a dedicated server in Tongji University. Nevertheless, to ensure the communication efficiency, the input and output of data is accomplished through the integrated DSRC (Fig. 9). The device of DSRC is equipped on roadside with solar panels (Fig. 10). Not sheltered by trees or lights, signals from traffic signal detectors and vehicles can be delivered to integrated DSRC and warning messages can also be sent to mobile terminals.

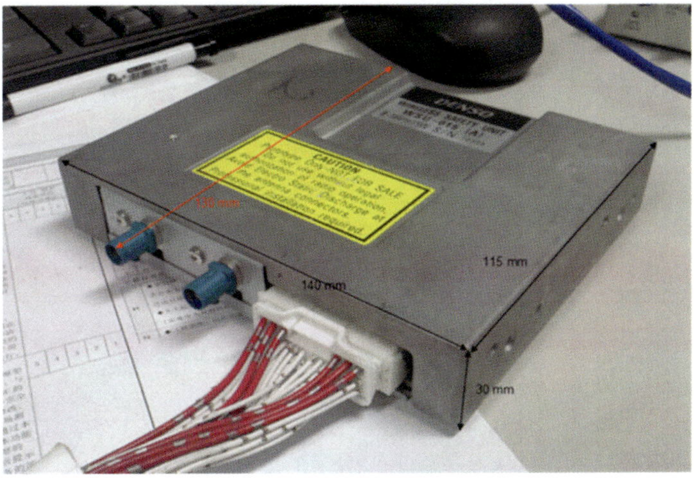

Fig. 9 Device of integrated DSRC

Fig. 10 Field investigation of DSRC

Table 4 Parameters of observed road at tested intersection

Parameters	Value
Design speed	60 km/h
Yellow interval	3 s
Width of the intersection	50 m
Average of observed speed near the intersection	40 km/h

A field observation was conducted at the signalized intersection of Gulang Rd.—Qilianshan Rd. Several parameters of the observed road (Qilianshan Rd.) are listed in Table 4.

The critical value of yellow interval is

$$\theta_0 = 1.84 + 5.04 = 6.88(\mathrm{s}) \tag{11}$$

For $\theta < \theta_0$, there would be a dilemma zone at upstream of Qilianshan Road. The boundaries of this area are

$$D_s = 20.64 + 24.64 = 45.28(\mathrm{m}) \tag{12}$$

$$D_0 = 24.64 + 33.6 - 56 = 2.24(\mathrm{m}) \tag{13}$$

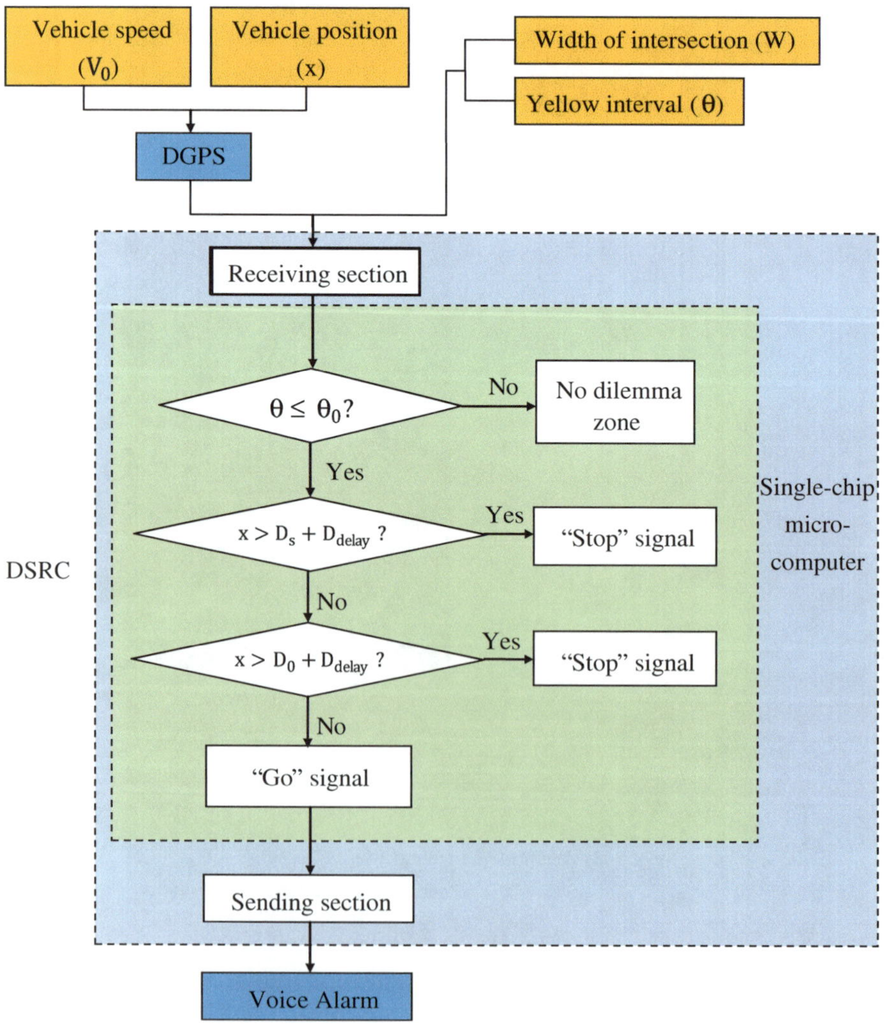

Fig. 11 Flowchart of LBS-based DZWS

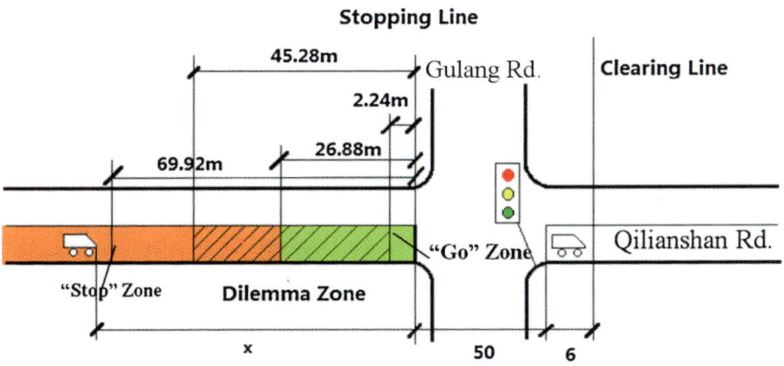

Fig. 12 Warning areas of the tested intersection (Gulang Rd.–Qilianshan Rd.)

Without the assistance of the warning system, the dilemma zone is 2.24–45.28 m away from stopping line. Besides, drivers would get alarmed of whether to stop or not 24.64 m ahead at the beginning of yellow interval (see Formula 14).

$$D_{delay} = 0.28 \times 40 \times T = 24.64(\mathrm{m}) \quad (14)$$

With the above value, the flowchart of LBS-based DZWS can be established in Fig. 11.

Through the analysis above, the voice alerter would alarm drivers 26.88–69.92 m ahead of intersection to "stop" and 0–26.88 m ahead of intersection to "go". The warning areas of the tested intersection are shown in Fig. 12.

According to this case, DZWS is feasible to prevent drivers from getting into dilemma zone in general condition, except for drunk driving or aggressive driving. The choice of 85th quantile of drivers' reaction time means that nearly 85 % of the tested drivers could take prompt action provided by voice alerter. This also ensures the system with 85 % accuracy.

4 Summary

In this research, warning areas and corresponding algorithm of yellow light dilemma zone have been presented. The establishment of the intelligent vehicle-road system—DZWS would prevent the behavior of running yellow light in a great deal and help drivers to drive safely near signalized intersections.

This driver-targeted system helps drivers to understand the degree of driving safety during yellow interval and to make reliable decisions with warning signals. With the help of in-vehicle devices and roadside units, drivers' wrong estimation of passing chance during yellow interval would be reduced if they abide by warning messages.

On the other hand, this system can scientifically help drivers to get rid of slam brake at intersections. In accordance with automobile dynamics, fuel consumption differs in varied speed and condition. For example, fuel consumption in 40 km/h is about 35 ml with one gentle brake but it would reach 100 ml with one slam brake. Hence, drivers would also gain profit in fuel cost with this system.

Future study would focus on the field tests of integrated DSRC and whole system in a larger scale in Shanghai and the targeted vehicles would cover trucks and buses. A more precise two-way communication system would be established, not only to warn the tested vehicles but also to alert other users (other cars, pedestrian and bicycles). Besides, a personalized algorithm of drivers' reaction time and self-adaptation system would be available with the accumulation of driving data.

Acknowledgments This research is supported by National High Technology Research and Development Program of China (863 Program)—Active traffic safety oriented urban vehicle location services online technology. The technical support of DSRC is provided by Wuhan University and Tongji University.

References

Cho CH, Su H, Chu YH et al. (2012) Smart moving: a SPaT-based advanced driving-assistance system. Network Operations and Management Symposium (APNOMS), 2012 14th Asia-Pacific. IEEE, 2012 pp 1–7

Gates TJ, Noyce DA, Laracuente L (2007) Analysis of dilemma zone driver behavior at signalized intersections, Transportation Research Board 86th Annual Meeting, Washington, DC

Heng W, Zhixia L. (2009) Dilemma zone modeling using yellow-onset vehicular trajectory data. In: Proceedings of the 9th International Conference of Chinese Transportation Professionals, ICCTP 2009: critical issues in transportation system planning, development, and management, vol 358, pp 886–894

Lei F, Lihua L, Rui T (2010) Performance of dilemma zone mitigation system at signalized intersections. In: Proceedings international conference on optoelectronics and image processing (ICOIP 2010), pp 113–15, doi:10.1109/ICOIP.2010.76

Li M, Boriboonsomsin K, Wu G et al (2009a) Traffic energy and emission reductions at signalized intersections: a study of the benefits of advanced driver information. Int J Intelligent Transp Syst Res 7(1):49–58

Li Y, He-wei Y, Zhong W (2009) The research of unintentional red running violation owing to dilemma zone, second international conference on intelligent computation technology and automation (ICICTA), 4:708–710

McCoy PT, Pesti G (2003) Improving dilemma-zone protection of advance detection with advance-warning flashers. Transp Res Rec 1844(1):11–17

Middleton D, Bonneson J, Hassan C (2011) New evidence of improvements in dilemma zone protection by detection-control system. Transp Res Rec 2011(2259):151–157. doi:10.3141/2259-14

Moon YJ, Lee J, Park Y (2003) System integration and field tests for developing in-vehicle dilemma zone warning system. Transp Res Rec 1826(1):53–59

Pant PD, Cheng Y (2001) Dilemma zone protection and signal coordination at closely-spaced high-speed intersections. No. FHWA/OH-2001/12. University of Cincinnati, Department of Civil and Environmental Engineering

Papaioannou P (2007) Driver behaviour, dilemma zone and safety effects at urban signalised intersections in Greece. Accid Anal Prev 39(1):147–158

Sharma A, Bullock DM, Peeta S (2007) Recasting dilemma zone design as a marginal cost-benefit problem. Transp Res Rec 2035(1):88–96

Sharma A, Bullock D, Peeta S (2011) Estimating dilemma zone hazard function at high speed isolated intersection. Transp Res Part C: Emerg Technol 19(3):400–412

Tong Z, Yu B, Yong-Hong Z et al (2009) A research of generalized yellow light dilemma zone and diver behavior for intersection collision avoidance system strategy, Intelligent Vehicles Symposium, 2009. IEEE, pp 924–928

Urbanik T, Koonce P (2007) The dilemma with dilemma zones. In: Proceedings, ITE district, p 6

Zegeer CV, Deen RC (1978) Green-extension systems at high-speed intersections. ITE J, 48 (11):19–24

ATSSS: An Active Traffic Safety Service System in Pudong New District, Shanghai, China

Hangbin Wu, Chun Liu, Junhua Wang, Lianbi Yao, Shuhang Zhang, Yi Li, Zhengning Li, Cheng Liu and Shouen Fang

Abstract In this paper, we report on an ongoing project supported by the Ministry of Science and Technology of China with Shanghai Pudong New District selected as the case area. The urban freeway pavement monitoring data from Pudong traffic police department and key vehicles data from a business company respectively have been collected to study traffic events, traffic accidents as well as the driving trajectory and status of key vehicles. Then by integrating the traffic data with the traffic safety warning model, the active safety service can be provided to other drivers through an iOS App. In this App, the functions of accident black spot alert service, traffic incident early warning service and key vehicles early warning service have been developed and included.

Keywords Active traffic safety · Data collection · Traffic events · Key vehicles · Early warning

1 Background

With the economic and social development, traffic accidents have become a major threat to the human beings, especially in urban areas. In recent years, more than 1.3 million people have been killed and over 200 million people injured due to traffic accidents worldwide. China is one of the countries which see the most road accidents in the world. The current road safety situation in China is very grim. In 2010, 3,906,164 road traffic accidents were reported, an increase of 35.9 % from the previous year. Among them, 219,521 cases were of casualties with 65,225 people killed and 254,075 injured. The direct property loss is more than 930 million RMB

H. Wu (✉) · C. Liu · L. Yao · S. Zhang · Y. Li · Z. Li · C. Liu
College of Surveying and Geo-Infomatics, Tongji University, Shanghai, China
e-mail: hb@tongji.edu.cn

J. Wang · S. Fang
School of Transportation Engineering, Tongji University, Shanghai, China

G. Gartner and H. Huang (eds.), *Progress in Location-Based Services 2014*,
Lecture Notes in Geoinformation and Cartography,
DOI 10.1007/978-3-319-11879-6_17

(according to Ministry of Public Security of People's Republic of China 2010). The problem of road accidents has become one of the main constraints on sustainable development of society and economy.

Based on the road accidents situation in recent years in China, it has been found that they are of the following characteristics:

(a) Due to frequent speed and lane track changes, urban roads and intersections and freeway interchange areas are places of the most concentrated road traffic accidents. In 2010, about 18.2 % traffic accidents happened in these places in China. In Shanghai Pudong New District, more than 60 % of the accidents occurred in the freeway entrances and exits. In these locations, illegal parking, lane change, reversing and overtaking are direct causes of accidents. If the vehicle behind can be informed of the speed and trajectory of the vehicle in front, the drivers can take proactive measures to avoid traffic accidents.
(b) Heavy casualties were mostly caused by key vehicles. Key vehicles include tourism chartered buses, class line coaches, vehicles with dangerous chemicals like explosives or fireworks, and dedicated civilian road vehicles. These vehicles are likely to cause death or serious injury in a road accident. Thus, if other vehicles could be informed of the location of key vehicles, some accidents would be avoided.

In summary, one of the main causes of traffic accidents is that drivers behind cannot obtain the traffic information of vehicles in front, and thus cannot make proactive measures in advance. If the location of relevant vehicles could be obtained in real time, proactive traffic safety intervention would be provided by analyzing the relationship between different vehicles and traffic events. In this circumstance, the active traffic safety system will play important roles.

The research of active traffic safety planning in America was started in late 20th century (Bishop 2000). Lots of design standards like Policy on Geometric Design of Highways and Streets (Aashto 2001) and Effect of Highway Standards on Safety (Warren and Daily 1995) and active anticollision system like Vehicle Infrastructure Integration (VII) (Misener and Shladover 2006), Cooperative Vehicle-Highway Automation System (CVHAS) (Shladover 2009) and IntelliDriveSM (Arnaout et al. 2010) were developed. These projects focused on the warning system based on the Vehicle to Vehicle (V2V) communication or Vehicle to Road (V2R) communication. Based on V2V and V2R, researchers in Europe and Japan also developed some active safety system like PreVent (Amditis et al. 2010), CarTalk2000 (Kosch et al. 2009; Reichardt et al. 2002), COMeSafety (Sukuvaara and Nurmi 2009), VICS (Tamura and Hirayama 1993) etc. Some of them have already been put into practice, forming an efficient traffic safety system. However, V2V and V2R technologies only hire relative position between vehicles for active warning model. The relative distance between vehicles were measured in real time and were transmitted to estimate the safety status. The absolute position of the vehicles were not selected for active traffic safety model. Therefore, some researchers pay their attentions to the real time position and trajectory of vehicle for active warning. D-GPS was mostly used for vehicles' position, speed and direction information acquisition for its high accuracy and high frequency (Morioka et al. 2000). Based on that, the traffic information such as real time traffic status could be used for warning systems. With the development of mobile phones, the active traffic safety model based on the real time traffic information is also necessary for a normal driver. In order to efficiently reduce the information gap between the drivers and provide road safety information in an appropriate way through an active traffic safety system, three problems must be solved. (1) the collection of basic data for the calculation of early warning information; (2) the design and presentation form of active traffic safety service; (3) the evaluation of the whole system and its application in other areas.

Location-based services technology can provide comprehensively integrated and personalized service by combining the real-time location information, spatial information and wireless communication technology. It can be effectively used for traffic safety services. In this paper, we report on an ongoing project supported by the Ministry of Science and Technology of China with Shanghai Pudong New District selected as the case area. The urban freeway pavement monitoring data and key vehicle data from Pudong traffic police department and from a business company respectively have been collected to study the traffic events, traffic accidents as well as driving trajectory and status of key vehicles. About 20 local intersections were selected as the black spot because the frequency of traffic accidents in these 20 intersections are significantly higher than other intersections. Then by integrating the traffic data with the traffic safety warning model, the active safety service can be provided to other drivers through an iOS App. In this App, the functions of accident black spot alert service, traffic events early warning service and key vehicle early warning service have been developed and included.

2 Active Traffic Safety Services System (ATSSS)

2.1 Case Area

The research area covered in this paper locates in the Pudong New District of Shanghai, P.R. China. In this area, the freeways are over 150 km including four tunnels and four bridges. According to previous research, over 60 % of the freeway accidents occurred at the entrances or exits of freeways. Therefore, in this study, we focus on the freeways of Pudong New District to collect traffic accidents and traffic events data and then provide relevant services to drivers in this area. The area within cyan lines in the following Fig. 1 shows the case area.

2.2 Main Research Prospective of This Paper

Compared with traditional anticollision systems, in this research, two more additional data sources, the real time traffic events data and the key vehicle data, were

Fig. 1 The case area

selected for an active traffic warning system. The real time traffic accidents data were shared from the police department, while the key vehicle data were transmitted from a company. These two kinds of data could be used as important data sources of active traffic safety model. Different with other anticollision system, the smart phone was selected as the terminal. At last, combining the two new kinds of data and the model, an iOS based App was developed and put into practice in Shanghai Pudong new district.

2.3 Design of the Active Traffic Services System

In this paper, an Active Traffic Safety Services System (ATSSS) is designed based on the video data of freeways and key vehicles data. This system consists of mainly two parts, (1) traffic safety database subsystem and (2) positioning and services subsystem. The design of active traffic services system is shown in Fig. 2.

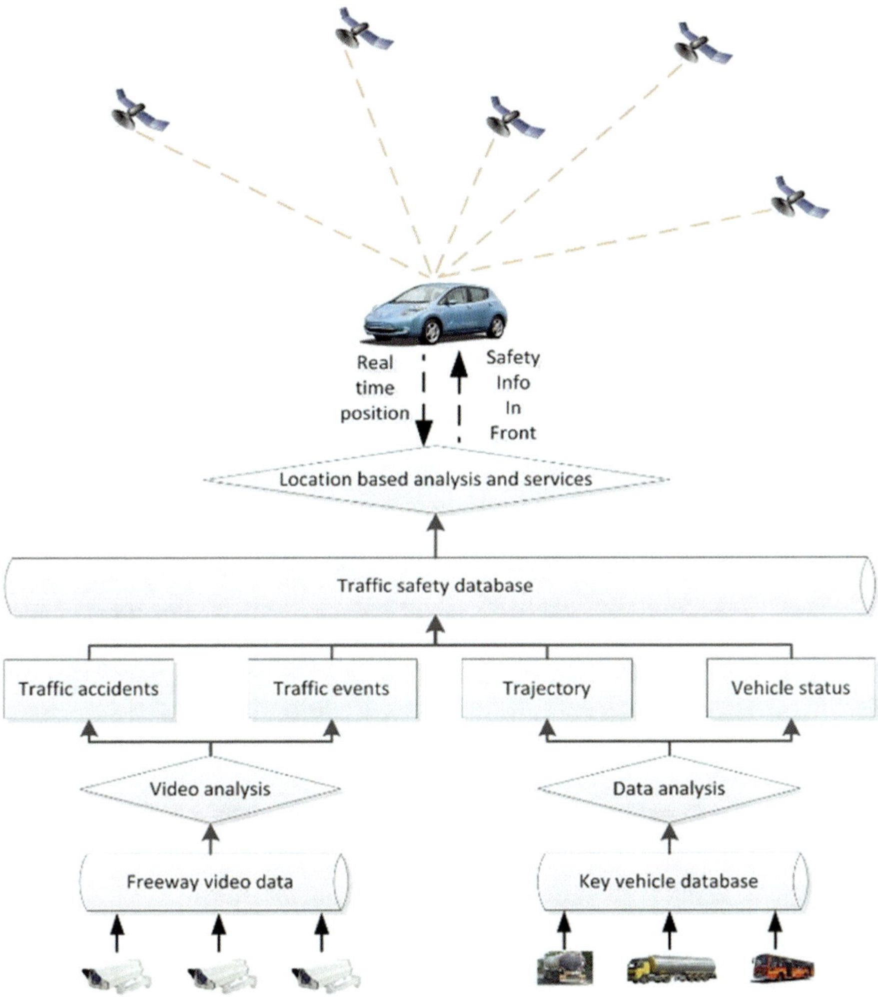

Fig. 2 The components of the proposed system

Traffic safety database subsystem mainly collects the traffic information of freeways and the status information of key vehicles. In this study, the traffic information of freeways is video data from the cameras installed along the freeways. After analysis, the traffic accidents and the traffic events data can be gathered.

The positioning and services subsystem collects the real-time positions of normal vehicles and then transmits them to the traffic safety database subsystem. After analysis, the necessary information such as illegal parking etc. can be delivered to the users.

3 Data Collection for Active Services System

3.1 Traffic Events Data Collection

Apart from roadside DV equipment and induction coil, a Traffic Event Data Collection Platform (TEDCP) is needed to integrate all the data collection devices, data receiving part and sending part. The diagram of TEDCP is shown in Fig. 3.

In Fig. 3, the blue part is the existing part installed in the traffic police department and the other part is established in our study. The purpose of the new part is to transmit the traffic events data detected and stored from police's server to our own server.

Two servers are used to send and monitor events data. One Data Transmission Server, (DTS) collects data from central database and uses wireless Data Transfer Unit (DTU) to send data to data collection server (DCS). To protect data security, a gateway is added between data transmission server and DTU. Fixed public IP or domain name keeps the process of data transmission always online.

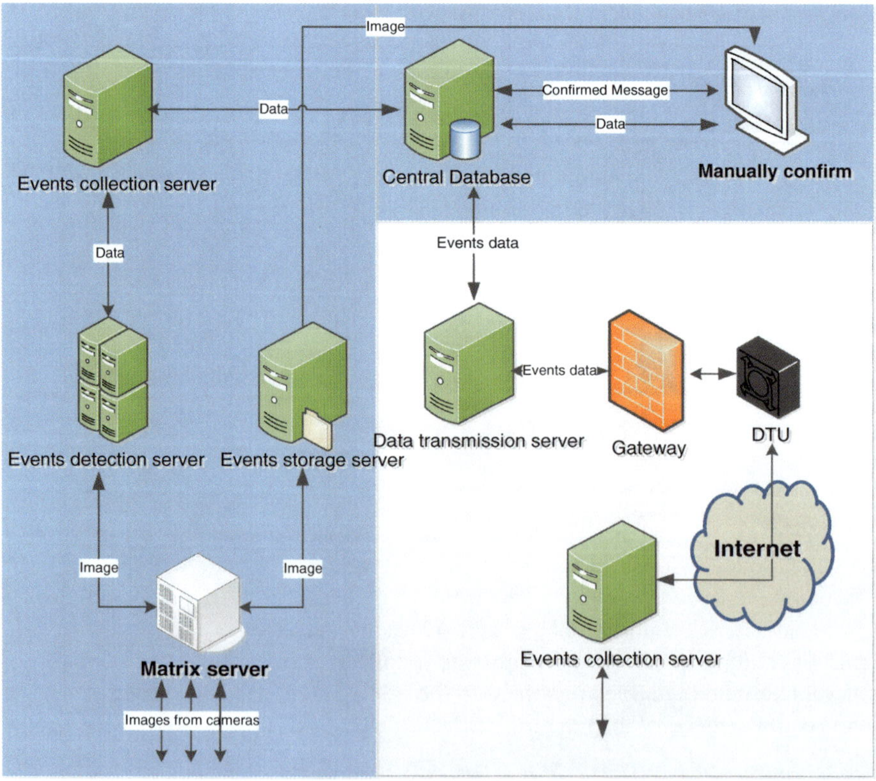

Fig. 3 Diagram of traffic event data collection platform (*TEDEP*)

Table 1 Data type and description of traffic event data collection platform

Name	Type	Description
GEID	NUMBER	
EVENTTYPE	VARCHAR2(10)	Type of event
POSX	FLOAT	X-coordinate of event place
POSY	FLOAT	Y-coordinate of event place
TIMEBEG	DATE	Beginning time of event
STATUS	NUMBER(5)	Current state of event
VIDEODATA	VARCHAR2(200)	Video path
EVENTSOURCE	VARCHAR2(100)	Event source (yhsd, jjvtd, citil)
CAMERAID	VARCHAR2(100)	DV number
LOCATION	VARCHAR2(20)	Event place
EVENTLEVEL	VARCHAR2(10)	Event grade

The way of data transmission is that the data collection progress which is running on DTS reads and codes traffic events, and uses RMQ to send the codes to data reception progress on DCS through gatekeeper.

In Pudong New District, with high speed communication network platform, data can be transmitted in real time. Administrators can have current traffic state in hand, thus improve transportation planning and management. To show the characteristics of traffic events, several data types (parking, reversing, turning around, speeding) are listed in Table 1.

3.2 Traffic Accidents Data Collection

Traffic accidents which occur on urban roads and freeways are mostly reported by traffic police. Traffic accidents data are collected by an iPad application developed for the convenience of traffic police. This increases the time effectiveness of dealing with a traffic accident. Data reported by traffic police, including time, location and other information relative to the accident, help to estimate the influence of the accident. Table 2 shows the specific information of each traffic accident.

3.3 Key Vehicles Data Collection

According to the Hazardous Article Transportation Management Ordinance, the hazardous articles include explosives, liquefied gas, radioactive substances and hazardous chemicals which must be transported by special vehicles. So the supervision of these vehicles becomes an urgent and important issue. In this study, the information of this kind of vehicles was shared by a company. The data of the

Table 2 Incident data to be reported

Name	Type	Description
Time	DATE	The time of the accident
Location	VARCHAR2(200)	The place that the accident took place
Weekday	NUMBER	The day that incident happened
Weather	NUMBER	Weather of the day
ReportingPeriod	NUMBER	Reporting time period of incident
Reporter	NUMBER	Reporter
ReachingSiteTime	NUMBER	Time spent by policeman to reach the site
IncidentType	NUMBER	Type of the incident
Blockedlanes	NUMBER	Lanes impacted to be blocked
FirstReachingBody	NUMBER	First rescue body that reached the site
LargeTrucksorBus	NUMBER	Number of large trucks or buses involved
VehiclesInvolved	NUMBER	Number of vehicles involved in the incident
PeopleEncage	NUMBER	Number of people encaged
Injuries	NUMBER	Number of injuries involved in the incident
Deaths	NUMBER	Number of deaths in the incident

Table 3 The fields of the key vehicle data

Name	Type	Description
TIME	VARCHAR	The time when the data is collected
LONGITUDE	FLOAT	The longitude of the vehicle
LATITUDE	FLOAT	The latitude of the vehicle
SPEED	FLOAT	The speed of the vehicle
STATUS	NUMBER	The status of the vehicle (on duty/off duty)
DIRECTION	NUMBER	The direction of the vehicle
ISLOCATE	NUMBER	To check the validity of the GPS

special vehicles are sent to our server via internet using the TCP/IP protocol, and a receiver program was developed according to the interface protocol. All the data were shared in real time. Each record we received includes 8 fields. Their descriptions are shown in Table 3.

3.4 Normal User Data Collection

Normal user data is gained by the GPS positioning in the mobile terminal. A location information database is designed in the server to store all the users ID and locations. The client sends location data to a server through fixed IP and the port every 15 s. The format of the uploaded data is shown in Table 4.

Table 4 Normal user data

Name	Type	Description
ID	String	The users identity
Time	String	The specific time of data acquisition
Longitude	String	Longitude of user's location
Latitude	String	Latitude of user's location

Table 5 Accident black spot data

Name	Type	Description
ID	String	The black spot's identity
UpdateTime	String	The specific time added to the table
Longitude	String	Longitude of black spot's location
Latitude	String	Latitude of black spot's location

The server could distinguish the latitude and longitude of different users at different times through user ID. Therefore, it can calculate and send related early-warning and service information to the users.

3.5 Accident Black Spot Data

In urban area, the accident frequencies of some roads' intersections are significantly higher than other intersections. In this research, we call this kind of intersection the accident black spot. According to previous research of Pudong traffic police department, 20 accident black spot areas are selected and several fields (see Table 5) are shared. In this paper, the fields are relative static.

4 Visualization and Service System

In this study, more data are included in order to provide safety services for normal users, such as road network, and accident black-spots. By combining all kinds of data with the traffic safety warning model, several map services and safety services were provided for normal users. In this section, map service and safety services will be introduced.

4.1 Traffic Safety Warning Model

To improve driving safety of important vehicles (large passenger cars, ambulances, fire trucks, dangerous goods vehicles and so on), a specific warning model is

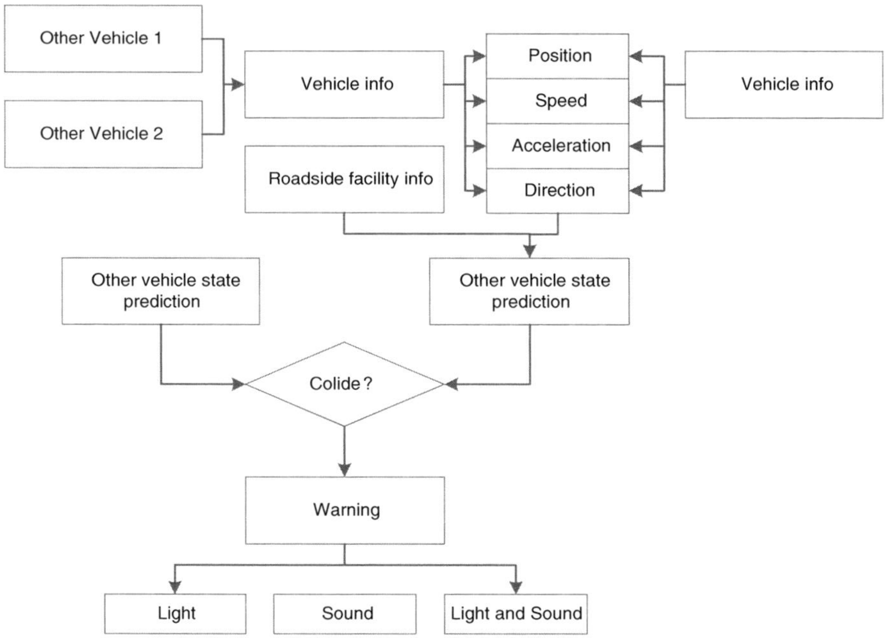

Fig. 4 Flowchart of traffic safety model

established. In this model, multimode positioning technology is used to collect vehicle state information which is transferred by wireless communication. Vehicle state prediction, including vehicle trajectory, is completed by Kalman filtering. If a collision is confirmed, different kinds of warning messages would be delivered to drivers who will then decide either to have an emergency stop or take proactive measures to avoid the other vehicle.

With collected information of other vehicles including key vehicles, own vehicles, traffic events and traffic accidents, drivers will calculate the gap or trajectory overlapping region between other vehicles and their own. Thresholds are selected in order to the trigger of warning section and the type of warning. The input and the output of the safety model is introduced in Fig. 4.

4.2 Map Service for Normal Users

This study uses Esri's ArcGIS software to complete digital map data processing and mapping. The publishing of map uses ArcGIS Server, and the development environment is ArcGIS Runtime SDK for iOS 10.1+XCode4.6. Operating system of Server is Windows 7 and mobile client iOS 6.1.

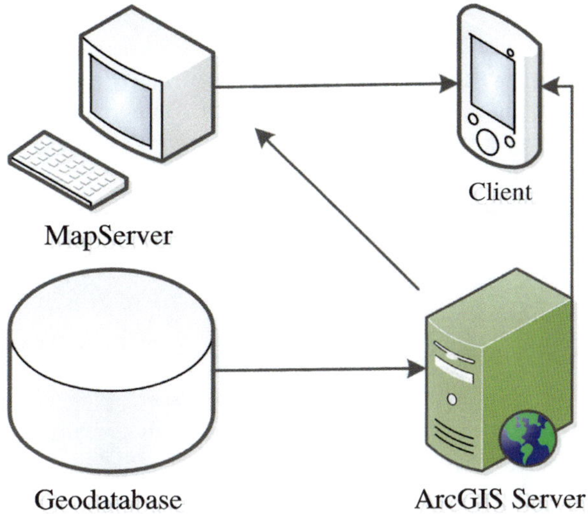

Fig. 5 The system structure

ArcGIS Runtime SDK for iOS is an application and development kit developed by Esri Company. It can be implemented on an Apple mobile device to access cloud platform providing the ability of GIS anytime and anywhere. It also can access and manipulate ArcGIS Online, ArcGIS Server and Portal for ArcGIS of GIS resources. The query, location, path analysis, and other complex geoprocessing tasks can be accomplished through interaction with the map elements. It shows great advantages in many ways, like real-time GPS positioning, intelligent traffic navigation and map view from multiple perspectives. It is an excellent platform to meet the needs of current and future mobile product development in the field of GIS and application service system development. Furthermore, it's a multi-application mobile development platform combining 3S and mobile computing in many areas. The system structure is shown in Fig. 5. Geodatabase is used for storing spatial data, which provides data sources for map services and path navigation. The system uses the data in Shapefile format to store geometric data. ArcGIS Server is used to load and run the server object and provide MapService and Network Analyst, which allows clients to replicate and extract data. Clients call on ArcGIS Server running MapService to generate a local map cache and find all the pictures within the display area based on the current display level and coordinate range. This different levels under 512×512 JPG format raster data map cache is transformed from the vector geographic data by ArcGIS Server Manager.

Functions of the mobile terminal include a call to Shanghai Representational State Transfer (REST) service, more details of the expression of the map and localization. These functions make it possible for subsequent path analysis, vehicle navigation and active Information Push Service.

The implementation of route analysis makes use of Route Task components of the ArcGIS for iOS API. Its data base is based on road network of Shanghai to generate a set of network dataset. We consider three kinds of conditions in the network dataset, namely the shortest length, the shortest time based on real time speed and the shortest time based on historical speed. At the same time we joined some constraints such as the ban, a one-way street, and turn. After drivers set a destination, the system can make route planning, and receive real-time traffic information from the server in the moving process in order to make automatic route re-planning. When the vehicle is about to travel to a particular stretch of road, such as road intersections, accident-prone sections and tunnels etc., the system can automatically switch to the detailed display view of that special section.

4.3 Active Traffic Safety Services

Active traffic safety services, which include accident black spot alert service, traffic incident early warning service and key vehicle early warning service, are integrated in the iOS system mobile client. Users can easily get access to these real-time safety services while driving. The safety service is overlaid in the navigation interface, including texts, images and incident points marked on the map. The alert is demonstrated in different ways—texts, images and voices, and the degree of urgency depend on the distance to the incident spot, the severity of incident and other factors.

- Text alert: shown on the map, the color changes according to the degree of urgency;
- Image alert: shown on the map, demonstrating the icon related to the incident;
- Voice alert: using Text-to-Speech (TTS) system, the text content and frequency depend on the degree of urgency.

- Accident black-spot alert service

Accident black-spot data are stored on the server. Mobile clients upload their current location constantly to the server. The client gets alert from the server when it reaches any of the accident black-spots. Table 6 shows the relationship between distance and warning voice, Fig. 6a is the accident black-spot alert UI of the mobile client.

Table 6 Accident black-spot alert text content

Distance (m)	Voice alert text content
1,000	Accident black-spot 1,000 m ahead, drive with caution
500	Accident black-spot 500 m ahead, drive with caution
200	Accident black-spot 200 m ahead, drive with caution
100	Accident black-spot 100 m ahead, slow down

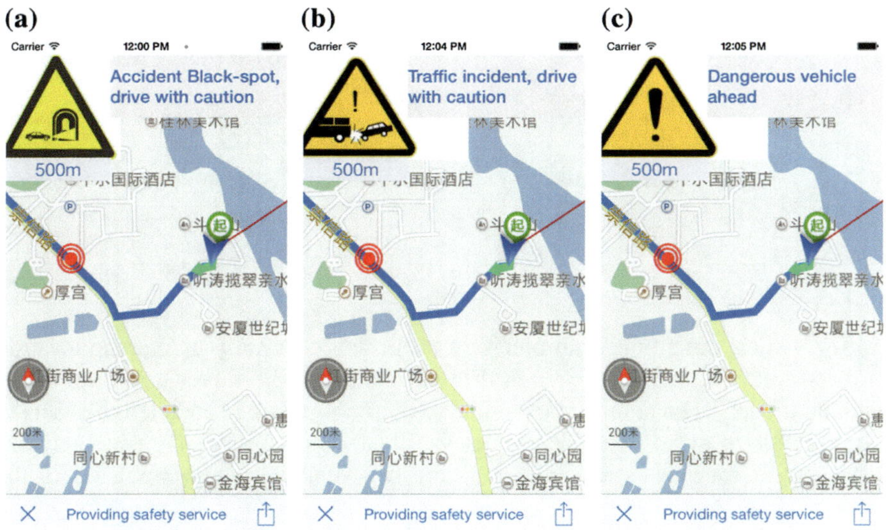

Fig. 6 Active traffic safety services for normal users through iPhone. **a** Accident black-spot alert. **b** Traffic incident early warning. **c** Key vehicle early warning

- Traffic incident early warning service

Traffic incidents include all kinds of traffic accidents such as rear-end, rollover, lateral collision, vehicle breakdown, spilled cargo and road work that blocks the lane. Traffic police reports an incident through the iPad application to the server. The influence area of the incident is then calculated by the server using the safety warning model. After that, the server informs clients in the influence area of the incident information. Then those clients judge whether to alert users according to whether the navigation route will cross the incident spot. Table 7 shows the relationship between distance and warning voice, Fig. 6b is the traffic incident alert UI of the mobile client.

Table 7 Traffic incident early warning text content

Distance (m)	Voice alert text content
2,000	Traffic incident 2,000 m ahead, pay attention to avoid
1,000	Traffic incident 1,000 m ahead, pay attention to avoid
500	Traffic incident 500 m ahead, pay attention to avoid
200	Traffic incident 200 m ahead, pay attention to avoid

Table 8 Key vehicle early warning text content

Distance (m)	Voice alert text content
500	Key vehicle 500 m around, pay attention to avoid
200	Key vehicle 200 m around, pay attention to avoid
100	Key vehicle 100 m around, pay attention to avoid

- Key vehicle early warning service

Key vehicle armed the high precision client will constantly upload the location to the server. The influence area is then calculated by the server using the safety warning model. Then clients in the influence area are informed and alert the users to pay attention to avoid those key vehicles. Table 8 shows the relationship between distance and warning voice, Fig. 6c is the key vehicle early warning UI of the mobile client.

5 Conclusions and Future Works

In this study, compared with the traditional anticollision system, two more kinds of traffic data have been used for the active traffic safety services. The traffic accidents and events acquired from the video were firstly used to make the traffic violation early warning for normal vehicle drivers. On the other hand, the driving trajectory and status information of key vehicles were delivered to normal vehicle drivers. By combining the two kinds of data and the relevant services, an active traffic safety service system has been established in Shanghai Pudong New District. For the convenience of normal users, an App based on the iOS system has been developed. This study puts traffic safety theories into practice and successfully forms a comprehensive active traffic safety system which is an important supplement of intelligent transport system.

In the near future, we will develop the system further and provide more specific and meaningful services to normal drivers, such as lane-conflict early warning and road construction early warning. These kinds of services depend on collected data, such as positioning results and road construction information as well as time delay of data transmission. Since the accuracy will significantly influence the results of services, some special mobile terminals are under development in order to obtain location with high accuracy and high frequency.

Acknowledgments This paper is supported by the National High Technology and Research Program of China (No. 2013AA12A206). Authors would appreciate the Pudong traffic police department and Zhongshan Changbao co., ltd. for their contribution to research data. Authors would also like to thank Miss Cao for her proofreading of this paper.

References

Aashto A (2001) Policy on geometric design of highways and streets. Am Assoc State highway and transportation Officials, Washington, DC 1:990

Amditis A, Bertolazzi E, Bimpas M, Biral F, Bosetti P, Da Lio M, Danielsson L, Gallione A, Lind H, Saroldi A, Sjogren A (2010) Holistic approach to the integration of safety applications: the INSAFES subject within the European framework programme 6 integrating project PREVENT. IEEE Trans Intell Transp Syst 11:554–566

Arnaout GM, Khasawneh MT, Zhang J, Bowling SR (2010) An IntelliDrive application for reducing traffic congestions using agent-based approach. In: Proceedings of the 2010 IEEE systems and information engineering design symposium, pp 221–224

Bishop R (2000) A survey of intelligent vehicle applications worldwide. In: Proceedings of the IEEE intelligent vehicles symposium, pp 25–30

Hughes WE, Daily K (1995) Effect of highway standards on safety. Transportation Research Board

Kosch T, Kulp I, Bechler M, Strassberger M, Weyl B, Lasowski R (2009) Communication architecture for cooperative systems in Europe. IEEE Commun Mag 47(5):116–125

Ministry of Public Security, P.R. China (2010) The road traffic accidents of 2010 in China. http://www.mps.gov.cn/n16/n1282/n3553/2921432.html

Misener JA, Shladover SE (2006) PATH investigations in vehicle-roadside cooperation and safety: a foundation for safety and vehicle-infrastructure integration research. In: Proceedings of the IEEE ITSC, pp 9–16

Morioka Yi, Sota T, Nakagawa M (2000) An anti-car collision system using gps and 5.8 GHz inter-vehicle communication at an off-sight intersection. In: IEEE VTS-Fall VTC, vol 5, pp 2019–2024

Reichardt D, Miglietta M, Moretti L, Morsink P, Schulz W (2002) CarTalk2000: safe and comfortable driving based upon inter-vehicle-communication. In: IEEE intelligent vehicle symposium, vol 2, pp 545–550

Shladover S (2009) Cooperative (rather than autonomous) vehicle-highway automation systems. IEEE Intell Transp Syst Mag 9:10–19

Sukuvaara T, Nurmi P (2009) Wireless traffic service platform for combined vehicle-to-vehicle and vehicle-to-infrastructure communications. IEEE Wirel Commun Mag 16(6):54–61

Tamura K, Hirayama M (1993) Toward realization of VICS-vehicle information and communications. In: Proceedings of the IEEE-IEE vehicle navigation and information systems conference, pp 72–77

Part V
General Aspects of LBS

Bridging the Gap Between Field- and Lab-Based User Studies for Location-Based Services

Ioannis Delikostidis, Holger Fritze, Thore Fechner and Christian Kray

Abstract There is a long-running debate about how to best evaluate mobile location-based services with users: in the lab or in the field? In this paper, we investigate how to combine benefits of both methods using an Immersive Video Environment (IVE), providing a convincing audio-visual simulation of real-world settings. We contrast three methods to evaluate mobile navigation systems: one in the field, one in the lab and one "hybrid" solution (IVE). We found that using the IVE allowed us to identify nearly the same number of major usability problems as the field test. We also observed similarities between the field study and the IVE study in terms of participants' performance, which provides initial evidence that in some settings, an IVE study may yield results comparable to a field study.

Keywords Usability evaluation · Mobile navigation systems · Immersive environments · Field-based testing · Lab-based testing

1 Introduction

Evaluating systems with users is an essential activity in the design and implementation process of mobile applications (apps), and in particular location-based mobile apps and systems. While it is not always useful to carry out (formal) user

I. Delikostidis · H. Fritze · T. Fechner · C. Kray (✉)
Institute for Geoinformatics (ifgi), University of Muenster,
Heisenbergstraße 2, 48149 Muenster, Germany
e-mail: c.kray@uni-muenster.de

I. Delikostidis
e-mail: dioannis@gmail.com

H. Fritze
e-mail: h.fritze@uni-muenster.de

T. Fechner
e-mail: t.fechner@uni-muenster.de

G. Gartner and H. Huang (eds.), *Progress in Location-Based Services 2014*,
Lecture Notes in Geoinformation and Cartography,
DOI 10.1007/978-3-319-11879-6_18

studies and not at all stages of development (Betiol and Cybis 2005), in many cases such studies play a central role in the evaluation strategies being adopted. There is a long-standing debate about what the best way is to evaluate a particular stationary or mobile system with users. Two fundamentally different evaluation approaches exist and researchers have argued strongly in favor (and against) either one: lab-based user studies ("in-the-lab"), taking place in the laboratory, and field studies ("in-the-field"), taking place in the real world. Both have specific benefits and drawbacks, e.g. in terms of repeatability, realism or cost/effort ratios. A key advantage of lab tests is the tight control over factors deemed relevant in the context of the experiment. A major benefit of field studies is a high degree of ecological validity. Key disadvantages of either method are the inverse of the other method's key benefits: lab-based studies usually lack a realistic context whereas field studies suffer from a lack of repeatability.

In this paper we take a fresh look at this discussion (Sect. 2) by comparing three different evaluation methods for mobile apps (Sect. 3) that make use of the user's location, such as mapping or navigation support apps. We carried out a field-based study, a classical lab-study and an evaluation using an Immersive Video Environment (IVE) (Sect. 4), and then compared the results obtained (Sect. 5). The aim of the study was to assess the differences between field and lab-based experiments and to evaluate the IVE as a means for bridging the gap between both methods. Our findings suggest that in some areas, using an IVE can lead to similar results to those obtained in field tests while maintaining most benefits of lab-based testing (Sect. 6). The paper concludes by summarizing the main contributions and pointing out directions for future research (Sect. 7).

2 Related Work

In mobile app development, there is a general trend towards context-aware, user-centered design and evaluation strategies (e.g. Lumsden 2008). A recent survey (Kjeldskov and Paay 2012) analyzed 144 publications from 2009 in the area of mobile HCI, and found a shift away from engineering-driven to empirical, evaluation-based research during the first decade of this century. They reported that in 2009, lab-based experiments appeared in 49 % of the total number of publications surveyed, while field studies accounted for 35 %. According to Kjeldskov and Paay (2012), most of those experiments were conducted in largely controlled rather than real usage settings. However, during the period they analyzed there has also been a steady increase of user studies with longer durations, carried out in natural usage environments (e.g. Chervest et al. 2002).

Undoubtedly, prototyping and testing in relevant, real contexts of use is on the rise (Rogers 2011), although the majority of mobile LBS research is still done in the lab (Lumsden 2008). Some researchers have argued that field-based usability studies are necessary to identify usability problems in depth (Nielsen et al. 2006). Or, that some behaviors could only be observed in the field (Duh et al. 2006) due to its high degree of realism (Jensen and Larsen 2008). People tend to behave more naturally (Kaikkoken et al. 2008) and the testing environment also corresponds to the one, where the tested system is meant to be used eventually. Unexpected behavior or appropriation is thus more likely to occur in the field.

Advocates of lab studies claim that public environments might make test users feel uncomfortable, distorting results (Palen et al. 2000): participants may behave "more negatively there than in the lab" (Duh et al. 2006). In addition, lab-based studies can be completed faster (Kaikkoken et al. 2008) and can hence cost less. The lab also provides reproducible and controllable conditions and has lower risks of interference or interruptions (Kaikkoken et al. 2008; Kjeldskov et al. 2004).

While the context of use has sometimes been considered too complex for realistic simulation in the lab (Kaikkoken et al. 2008), it is in principle possible to bridge the gap between lab- and field-based experiments by simulating the real world. Researchers can create a life-like, immersive environment, experienced by participants (Loomis et al. 1999). This type of experiments was initially used in (cognitive) psychology (van Veen et al. 1998) but has also been applied in mobile HCI research (Singh et al. 2006; Snowdon and Kray 2009; Ostkamp and Kray 2014). Such simulated environments can be realized in different ways to create experiences that vary greatly in (visual) fidelity. Entirely synthetic simulations (e.g. using textured 3D-models) enable free movement but are often lacking in terms of realism, whereas video-based systems limit motion while providing realistic visuals.

A range of usability testing methods for mobile interfaces, such as audio/video recording (van Elzakker et al. 2008) or screen-capture functionalities built into modern smartphones (Jensen and Larsen 2008) can be applied both in the lab and in the field. There are many further methods for field-based usability evaluations (e.g. experience-sampling, post hoc interviews) but yet no standardized set of methods exists (Lumsden 2008). To improve field-based data collection quality, different methodologies are used such as automatic logging or "lab-in-a-box" (Kimber et al. 2005; Winters et al. 2001), compact mobile usability labs (Betiol and Cybis 2005; Kaikkoken et al. 2008) or multi-camera mobile usability labs (Delikostidis and Van Elzakker 2009; Roto et al. 2004). The latter is seen by many researchers (e.g. Delikostidis and van Elzakker 2009; Goodman et al. 2004; Roto et al. 2004) as a potential solution for both lab and field experiments. In our research we therefore relied on this method to carry out our comparison study.

Overall, we can conclude that while many methods exist for the evaluation of location-based services, there is still a need to more precisely assess their relative benefits, beyond their basic characteristics. In the remainder of this paper, we therefore present a comparison study, where three methods were applied to the same scenario/system in order to gain a deeper understanding of their relative strengths and weaknesses.

3 Three Evaluation Settings

For our study, we applied the same evaluation techniques in three different environments: a real city setting (field test), a (non-realistic) lab-based setting (classic lab test), and a "hybrid" solution using an Immersive Video Environment (IVE). In the **field test** participants used a typical LBS application—Google Maps (http://www.google.com/mobile/maps/)—in situ, running on a GPS-equipped Android smartphone. We used a multi-camera recording system as for collecting data during uses, which also facilitated the use of the Think-Aloud method for data collection. Figure 1 (right) shows a person wearing the lightweight recording equipment.

The **classic lab-based setup** is depicted in Fig. 1 (left). Participants sat in front of a desktop screen, which showed high-resolution panoramic photos of actual locations while they interacted with Google Maps on a smartphone. A GPS simulation at OS level enabled us to synchronize photos and GPS positions. When a photo of a location was shown, the GPS position of the smartphone was set to the location where the image was taken. In principle, the same recording technique as in the field test was used except for participants not having to carry a backpack with the recording equipment. Instead, it was installed in an enclosure behind the PC screen.

For the second lab-based study our goal was to build a setup with a high degree of realism. Using panoramic video footage of real locations, we created an **IVE** similar to the one proposed by Singh et al. (2006) and Snowdon and Kray (2009). It consisted of three back-projection screens, arranged in a roughly semicircular way displaying panoramic video scenes, and a surround-sound system (cf. Fig. 1, middle). Each screen measured 200 × 150 cm, with a resolution of 1,280 × 1,024 pixels, providing an overall resolution of 3,840 × 1,024 pixels across all three screens. The angle between central and left/right screens was about 110 degrees. We implemented a custom software stack to control playback and select which video to show in the IVE. A standard database stored the videos, captured at the same real-world locations used for the field-based tests, along with their geographic coordinates. In the resulting system, when a scene is selected in the control app, the corresponding footage is played back in the IVE. Simultaneously, the location of an attached smartphone changes to the displayed scene's location using the GPS simulation as described above. The user testing methods for the IVE study was the

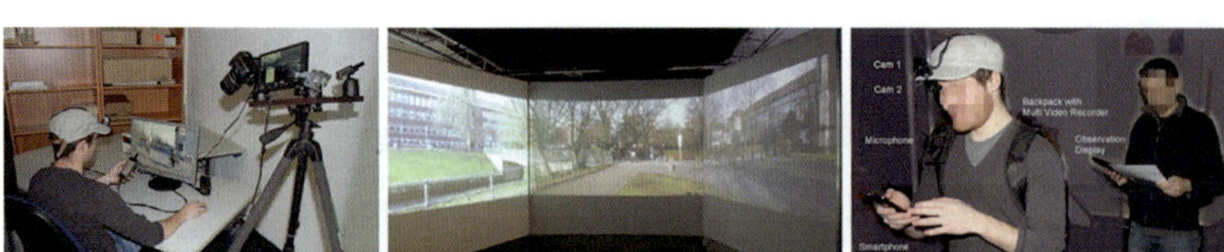

Fig. 1 Three evaluation setups: classical lab-based study (*left*), immersive video environment (*middle*), and field study (*right*). The *right hand* photo shows a participant wearing the multi-camera recording equipment and the experimenter in the back

same as in the other two settings: a combination of observation, think-aloud and audio/video recording, using the same multi-camera system worn by participants during the field study.

4 Comparison Study

The overall aim of our study was to investigate the (relative) advantages and disadvantages of IVEs compared to standard lab-based and field studies. A secondary goal was to identify relevant properties of each method that could inform the selection of evaluation methods for LBS.

Participants We recruited 18 participants (six for each type of study) via advertising through university channels, social networks and poster ads. People interested in participating were sent a pre-selection questionnaire asking about their background, familiarity with the study area, orientation capabilities, and previous experience with smartphones and navigation apps. All participants had some experience with smartphones, some familiarity with the study area, and most were students at a local university.

Material and Apparatus For our study, we used Google Maps as a freely available map-based LBS with route planning and guidance functionalities. Participants had to perform a series of 9 tasks, each consisting of 2–5 sub-tasks, given to them in printed form. The same tasks (cf. Table 1 for an example) were used for

Table 1 Example scenario and its task description (task 1)

Task	Course of action
1. Arriving at starting point and self-orienting at unfamiliar city location	Initial orientation/information searching/ destination finding
2. Crossroad re-orientation and finding post office	Route calculation/route verification/direction estimation
3. Crossroad re-orientation and finding meeting point	Route calculation/route verification/direction estimation
4. Crossroad re-orientation and finding important landmark	Information search on map/route verification/direction estimation
5. Crossroad re-orientation and discovering hidden pedestrian path	Changing map type/route verification/ direction estimation
6. Arriving at meeting point and finding final destination	Destination verification/route calculation/ direction estimation
7. Crossroad re-orientation and finding closest museum	Information searching/route verification/ direction estimation
8. Crossroad re-orientation and finding four neighboring POIs	Information search on map/route verification/direction estimation
9. Arriving at destination and self-orienting	Destination verification/direction estimation

Table 2 Overview of actions needed to complete each task

Scenario	Task description
Your friend gives you a ride until the parking of his work somewhere in Muenster and you need to find where you are and how to go to your destination. In doing so you check what exists around you (the closest bank)	(a) What is the name of the street in front of you?
	(b) Please find the closest Sparkasse bank
	(c) To which direction you think it is?
	(d) Please find your final destination (post office in Domplatz) using the search function
	(e) To which is your final destination?

all three methods. Each task was based on a short scenario, to be read before going through and executing the sub-tasks with the help of Google Maps (running on a provided LG Optimus P990 Android-based smartphone). For each task to be completed successfully, a specific course of actions was required (see Table 2).

For capturing the panoramic video footage shown in the IVE, we used three DLSR cameras (Canon EOS 550D), mounted on a custom-made adjustable bracket attached to a tripod. The angle between the cameras corresponds to the angle between the three IVE screens, on which the videos were later played back. Audio was recorded using a portable high-quality surround recorder (Zoom H2n). The recorded footage was post-processed with standard video editing software to minimize seams and overlaps, adjust for the resolution of the IVE, and to extract the panoramic photos used in the classic lab-test. In total, we created 9 video scenes and panoramic photos.

The mobile recording system (cf. also Fig. 1—right) was developed by Delikostidis and van Elzakker (2009) and consisted of three high-resolution mini cameras: two of them installed on a hat worn by the participant and one worn by the experimenter. It also included a 4-channel compact A/V quad (synchronized) recorder, an observer display, a battery, an HDMI-to-composite video converter (for capturing the smartphone screen), and a microphone. A 10 m-multicable transferred video signals to and from the participant. The system enabled us to synchronously record the screen content, what was being said, as well as the immediate and further environments from the participants' and the experimenter's perspective.

Ensuring Comparability To ensure comparability between the three evaluation methods, we applied a number of measures. Firstly, the same multi-camera recording mechanism was used in all three settings, resulting in the same raw data for each method. Secondly, the same locations were used in all three cases. This eliminated effects that could have resulted from different locations and entailed a between-group design for our study. We also recorded panoramic video footage at the exact locations, where participants were placed in the field tests. The goal here was to reduce the impact of different perspectives or point of views on the overall results. Finally, we selected tasks that could be executed in all three environments (e.g. no tasks required participants to move while interacting with the smartphone).

This measure was necessary to avoid influences resulting from certain tasks being very difficult to perform in one of the environments. We did however not try to fully control any unanticipated or environmental factors but aimed instead at observing those factors and their possible influence.

Pilot Testing To make certain that the experimental design was feasible and to calculate the required task completion times, we carried pilot tests for each setting. These let do a number of adjustments. One of those was the use of higher-capacity batteries to not risk possible power shortages during test sessions. The pilot test also brought to light the need to verbally describe the path between two simulated locations in the IVE to avoid disorientation. We also discovered that the search history of Google Maps had to be clear prior to each trial to avoid unwanted auto-completion.

Procedure At the start, participants received a description of the aims of the study and instructions for correctly carrying out the given tasks. They then were handed a smartphone running Google Maps, and were informed about which functions they could use during the study. They were also shown how to complete some example tasks, and were then given time to get familiar with the app. The time reserved for the briefing of all three experiments was 20 min (five for reading the description, five for demonstrating Google Maps and ten for getting familiar with the app). All the briefing sessions were held indoors before starting the test sessions.

Each field-based test session began by transporting the participant to the start location by car, and by assisting them in putting on the recording equipment. Once participants were ready and the recording system was operational, they were handed the smartphone with Google Maps, and a list of the scenarios and tasks. They were then instructed to read the first scenario and carry out the described sub-tasks. When a task was completed or abandoned, the researcher and the participant walked to the next location. Once there, participants had to read the next scenario. A very similar procedure was followed in the classic lab test and the IVE. However, instead of taking people to the start location, they were seated at the test desk (in classic lab test) or set standing in the middle of the IVE. When the recording equipment was ready, the photos or footages playback started. In the lab, when a photo or video of a new location was shown, participants were provided with a verbal description of the direction from which they would have arrived at the current location. This direction was also indicated on the footage. The aim was to help them imagine the process of traveling between two locations and re-orient themselves at the new location. After completing all tasks, participants were debriefed and asked to fill in a post-session online questionnaire assessing usability of Google Maps in terms of its usefulness, ease of use and satisfaction (based on the USE questionnaire by Lund 2008). All questions were answered using a 5-point Likert scale. Participants received a small payment for taking part in the study.

Analysis When we analyzed the data collected during the study—in the form of video files with audio (cf. Fig. 2) and questionnaire data—we measured performance

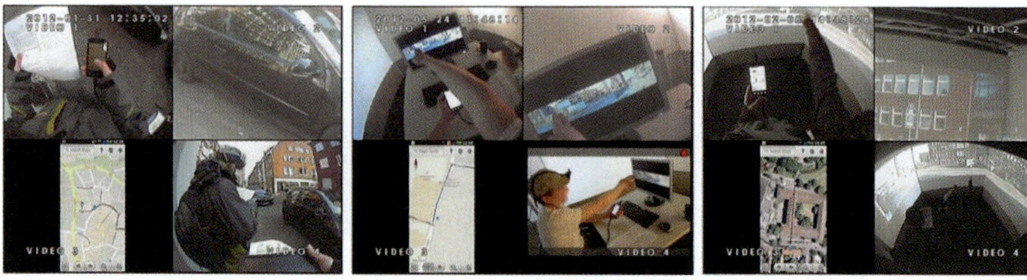

Fig. 2 Material recorded in the field (*left*), in the lab (*middle*) and in the IVE (*right*). *Top left* and *right quarters* show images acquired from two hat-mount cameras, *bottom left quarter* shows screen content of smartphone and *bottom right quarter* an external view of the TP

in terms of effectiveness and efficiency. Effectiveness, efficiency and number of usability problems were directly measured by reviewing the video recordings. Two researchers transcribed and analyzed the footage, and their results were compared for cross-validation. The average inter-rater difference for task completion time was 2.6 %; results for task completion (success) were identical. For the final analysis we used the average of the times identified by the raters. The number of usability problems was calculated based on a combination of participants' negative comments (as recorded through thinking-aloud), unnecessarily repeated actions, long pauses during the use of particular functions (defined as confusion points), and the inability to use particular functions. We extracted further usability problems from the online questionnaire, which also helped to assess usefulness, ease of use and satisfaction. Along with transcription, the amount of resources required for each of the methods was compared by calculating the total time spent for each test session. This was extracted from the contained on-screen time/date information of the video material, and from notes taken during the experiments. Due to the focus on qualitative aspects (e.g. usability problems) and the goal to get an initial and broad impression of relative differences and similarities of the three methods, we did only recruit 18 participants in total (6 per method) and thus did not obtain enough data points to carry out a meaningful inferential analysis.

5 Results

The total length of the video footage (cf. Fig. 2) was ca. 11 h, with an average of 37 min per participant. All the data was used in the analysis.

The average **effectiveness** in task execution was calculated as the proportion of successfully completed sub-tasks of each task over the total number of sub-tasks (25). Partially completed sub-tasks were assigned a coefficient of 0.5 instead of 1 (when fully successful). This occurred, e.g. when an estimate of direction was inaccurate but still within the range of ±45 degrees from the correct direction. Sub-tasks were considered unsuccessful, getting a coefficient of 0, when test persons either provided a wrong answer, or executed an incorrect action, or exceeded the

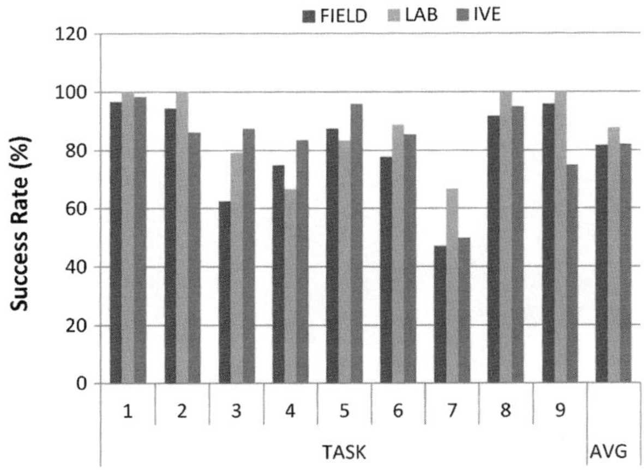

Fig. 3 Average success rates per task for each method (cf. Table 2 for task descriptions)

maximum time allowed for task completion (set to 4 min. based on pilot test results). The success rates for each task in each test setting are shown in Fig. 3. In this diagram a weight value was given to each task, calculated by dividing the number of its sub-tasks by the number 5 (the maximum number of sub-task per task).

The average weighted effectiveness for all tasks after normalization was 81.7 % for the field, 82.0 % for IVE, and 87.6 % for classic lab test. The results are nearly identical for the field and IVE tests and noticeably worse than classic lab test. During the latter, 4 out of 9 tasks were completed successfully (Tasks 1, 2, 8, and 9). For these particular ones, neither the field nor IVE reached that score. However, in one of the two tasks with the lower score for classic lab test (Task 4), both field and IVE produced better results. That task coped with route verification and orientation using Google Maps, localizing a specific local landmark and estimating its direction. The lowest overall score achieved in Task 7 in the field, which involved searching for the closest museum using Google Maps and estimating its direction.

Figure 4 summarizes the measured average **efficiency** in terms of time for successfully completing a task. The tasks that were not successfully completed were not taken into account for the calculations. From Fig. 4 we can note a number of observations: Task 7 was not considered, as only one field test participant (TP4) succeeded on it, no one in the IVE, and only two (TP13 and TP16) in the lab study. The latter produced the fastest completion times in all but one task, and the field study had the fastest time in task 8. IVE participants took the most time to complete 6 out of the 8 tasks—in three cases by a large margin (task 6, 8 and 9).

The average scores given by participants in the online questionnaire in the categories **usefulness, ease of use** and **satisfaction** were higher for the classic lab test compared to the other two methods. Interestingly, the IVE produced better overall results than the field test. The scores were (on a five-point Likert scale; higher values are better) as follows: Usefulness: field: 3.33; IVE: 3.93 and classic lab 4.27. Ease of use: field: 3.63; IVE: 3.67 and classic lab: 4.30. User satisfaction: field: 3.27; IVE: 3.63 and classic lab: 3.73. Figure 5 summarizes the ratings for each evaluation method.

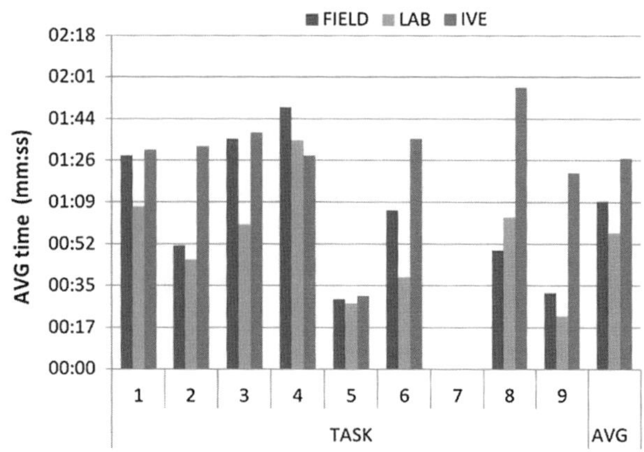

Fig. 4 Average (AVG) times per task for each method (cf. Table 2 for task descriptions)

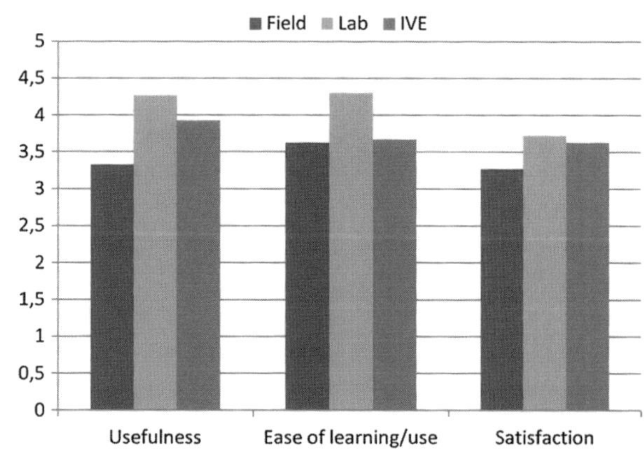

Fig. 5 Average ratings for three evaluation methods (from post-study questionnaire)

We categorized the **usability problems** identified during the three experiments, in terms of their severity, as either minor or major ones. Minor problems were predominately cosmetic, e.g. interface color schemes or other easily rectified issues. Major issues were severely affecting a user's ability to correctly complete tasks. An example for this was the persistence of Google Maps to draw a route from an earlier position, although the user had already moved to a new place. This recurrently caused frustration and confusion, making user orientation difficult. The total number of usability problems found was 23: 22 during the field tests, 16 in the classic lab test and 17 in the IVE. Ten major problems were found, most of which (nine) in the field-tests, seven in the lab-test and eight in the IVE. Thirteen minor problems were discovered in the field, and nine in each the classic lab-test and the IVE.

Cost and effort of evaluation methods are rarely discussed but nevertheless important. We thus report on our experiences in terms of resources and effort spent when applying the three methods. To do so, we calculated the average time spent for executing each experiment after completing the tests with all 18 participants.

The total time needed per participant in the field (79 min) was considerably more than two other methods (40 min for classic lab setup, 41 for IVE). Having to physically move between locations instead of almost instantly being 'teleported' there clearly contributed to this difference, as did the need to transport participants to the test area. Preparations and debriefing also took longer in the field, e.g. to change battery packs. Since we collected exactly the same data in all three tests, the data analysis time was identical.

In addition to time spent per participant/session, preparing a study also takes time, harder to quantify. Locations have to be scouted, and in the case of IVE and classic lab test, panoramic footage has to be captured and post-processed. The estimated time for capturing footage per location is less than 10 min, and for post-processing per site less than 15 min. Custom software could speed up post-processing considerably. Additionally, footage has to be tagged to be incorporated in the test software and links to other footage have to be specified—this step takes less than 5 min.

In terms of cost, there were a number of factors constant across all three methods: participant fees, recording equipment, data storage and analysis, questionnaire printing, etc. As we used existing equipment, we can only estimate the cost for setting up an IVE, consisting of three short-throw projectors plus projection screens (ca. €4,300) and a surround-sound system (ca. €200). The tri-camera system cost approximately €1,900, and the high-end PC driving the system set us back ca. €700. In total, there were thus one-time setup costs of ca. €7,100. This amount of course only constitutes an estimate based on the components used at the time of the study.

We also spent ca. 80 man-hours building the control software for the IVE. Overall, we can therefore observe that while there were considerable one-time costs for setting up an IVE, studies can be run in less time and thus at a lower cost. For instance, if footage is re-used, further cost/effort benefits can be obtained. Furthermore, the calculated costs are based on our three-projector IVE setup. Using alternative approaches e.g. a single High-Definition projector, could reduce costs even more.

6 Discussion

The comparison study provided us with some initial insights into the relative benefits and drawbacks of each evaluation method. The field test identified the highest number of major **usability problems**, followed by the IVE (9, resp. 8 problems) and the classic lab test (7 major problems identified). Out of the total 23 usability problems identified, 7 were major and strongly context-related: 6 of those were found in the field, 5 in the lab and 6 in the IVE. One of them (an important landmark, i.e. a tall church clearly visible in reality but missing from the map) was found only in the IVE.

The **task effectiveness** rate was highest in the classic lab test (87.6 %) and almost the same for field and IVE tests (81.7 and 82 %). This could be due to the lack of distraction from the environment in the (classic) lab, which participants were exposed to in the field and in the IVE (e.g. visual input and movement). The similar results for field- and IVE tests hint at participants exhibiting similar behavior and strategies in both settings. IVEs therefore show promise to recreate (certain) aspects and contextual factors of reality with experimental results comparable to field tests.

Generally, the IVE produced the worst **task efficiency** results but we can also observe that field and IVE tests produced identical or very similar results (different from the classic lab test) in six out of eight tasks. The limited projector resolution might have contributed to this result, as it reduced the legibility relevant elements such as street name signs. The high-resolution photos in the classic lab study did not suffer this problem. In context-sensitive tasks (e.g. task 5, which involved re-orientation with a totally different map type), the results of IVE and field tests were nearly identical or at least had the same trend compared to the lab test results.

Obviously, our study was subject to a number of **limitations**. The number of participants was quite small, as the study was of qualitative rather than quantitative nature. Since a key goal was to analyze which usability problems could be found per condition, a between-group design was more feasible. A within-subjects design could have been alternatively used, however carryover effects would be strong in the particular experiment and difficult to compensate, e.g. with respect location(s) used in the study. According to Nielsen (1994) a minimum of five participants is sufficient to uncover most usability problems without wasting resources. A larger number would be preferable if participant groups with very specific characteristics were tested. In our case however, the participant characteristics of the groups were largely comparable and the system interface under test was the same. Secondly, unusually cold temperatures affected the mobility of participants during the field study due to the thick layers of clothes they wore. Using a smartphone while wearing winter gloves is cumbersome. The field-study required a larger physical effort compared to the other two methods, which we controlled for by using lightweight equipment. A further limitation relates to locomotion, which was not possible in the lab test and within the IVE. To ensure comparability we thus chose tasks, where participants remained stationary. While people interact with their mobile phone while walking, earlier studies (e.g. Duh et al. 2006) have shown that they in fact often stop and/or seek quieter places to do so. For this study, the focus was on interaction with mobile LBS in different environmental settings and not on interaction while moving. Finally, classic lab and IVE tests only provided 180° photo and video panoramas, while field tests facilitate the full 360°. Table 3 summarizes key benefits and drawbacks of each method.

Several of the limitations mentioned above could be addressed in future experimental setups. Using HD-projectors would address the lack of resolution and a 360° environment could be constructed. To enable locomotion, a treadmill or similar tools could be used (e.g. Schellenbach et al. 2009). However, since video footage does not support free exploration of the recorded area (unlike synthetic 3D worlds), movement would still have to be restricted and/or large amounts of video

Table 3 Comparison of pros and cons for the three methods

	Pros	Cons
Field	Most usability problems found	Largely dependent on weather conditions
	Testing in the target environment of LBS	Often more difficult to find participants (weather conditions influence)
	Running costs stable in the long term	Longer duration per participant
		Transportation to test areas costs time and money
Lab	Simple setup	Contextual factors almost absent
	Requires little space for the setup in the lab	Time needed for panoramic photo collection
	If footage re-used, costs are reduced considerably	Performance results different from real world tests
	Relatively low cost	Locomotion not possible
		Some test tasks need adjustment to setup limitations
IVE	Several contextual factors can be recreated effectively	Some contextual factors complicated to recreate
	Number of identified major usability problems approximates that of field tests	Time needed for video footage collection
	Large automation possibilities	Locomotion possible but costly or unnatural
	Considerable cost benefits in the longer term	Requires more space for the setup in the lab
	If footage re-used, costs are reduced considerably	Test tasks have to be adjusted to setup limitations
		Larger one-time costs

footage at different walking speeds would have to be collected. While different temperatures could also be simulated inside IVE, this is less true for other weather conditions e.g. precipitation. The unnatural accuracy of the simulated GPS position compared to the (fluctuating) accuracy of GPS receivers outdoors also deserves further attention. As the latter largely affects the use and usability of mobile LBS, GPS error simulation (e.g. based on recorded GPS data) in the lab during usability testing experiments would produce more realistic results.

7 Conclusion

In this paper we presented a comparison study between a field-based and two lab-based setups for evaluating mobile LBS. The results indicate that Immersive Video Environments (IVE) show potential as an evaluation method that combines benefits of both field- and lab-based studies (e.g. repeatability, control, and rich contextual information). In many cases (e.g. number of major usability problems found, effectiveness, task efficiency), similar results were obtained in the IVE and in the field. However, some differences exist and need to be considered when planning a user study. In particular, the number of (minor) usability problems identified was higher in the field test, and the lack of locomotion in the IVE and classic lab study can cause issues for certain types of apps (e.g. those meant to be used while walking). The results reported here only provide initial insights into the benefits and drawbacks of the three types of evaluation methods but further research is needed to fully understand when to pick which method and what the implications are of choosing each one. In addition to carrying out further comparison studies (with different applications and evaluation methods), there are also several promising options of extending the IVE approach to overcome issues such as disorientation or lack of locomotion.

References

Betiol A, Cybis AW (2005) Usability testing of mobile devices: a comparison of three approaches. In: Proceedings of interact 2005. Springer, Berlin, pp 470–481

Chervest K, Mitchell K, Davies N (2002) The role of adaptive hypermedia in a context-aware tourist guide. Commun ACM-The Adapt Web 45(5):47–51

Delikostidis I, van Elzakker CPJM (2009) Geo-identification and pedestrian navigation with geo-mobile applications: how do users proceed? In: Proceedings of the 5th international conference on location based services and telecartography. Springer, Berlin, pp 185–206

Duh HBL, Tan CBG, Chen VHH (2006) Usability evaluation for mobile device: a comparison of laboratory and field tests. In: Proceedings of mobile HCI '06. ACM, New York, pp 181–186

Goodman J, Brewster S, Gray P (2004) Using field experiments to evaluate mobile guides. In: Proceedings of HCI in mobile guides, workshop at mobile HCI 2004. Springer, Glaskow, pp 38–48

Kaikkoken A, Kekäläinen A, Cankar M, Kallio T, Kankainen A (2008) Will laboratory test results be valid in mobile contexts? In: Lumsden J (ed) Handbook of research on user interface design and evaluation for mobile technology. Information Science Reference, Hersey, pp 897–909

Kimber J, Georgievski M, Sharda N (2005) Developing usability testing systems and procedures for mobile tourism services. In: Proceedings of the annual conference on information technology in the hospitality industry, HITA 2005. Los Angeles, USA, pp 79–96

Kjeldskov J, Paay J (2012) A longitudinal review of mobile HCI research methods. In: Proceedings of Mobile HCI '12. ACM Press, San Francisco, pp 69–78

Kjeldskov J, Skov MB, Als BS, Høegh RT (2004) Is it worth the hassle? Exploring the added value of evaluating the usability of context-aware mobile systems in the field. In: Proceedings of mobile HCI '04. Springer, Berlin, pp 61–73

Loomis JM, Blascovich JJ, Beall AC (1999) Immersive virtual environment technology as a basic research tool in psychology. Behav Res Methods Instrum Comput 31(4):557–564

Jensen KL, Larsen LB (2008) The challenges of evaluating the mobile and ubiquitous user experience. In: Proceedings of 2nd international workshop on improved mobile user experience, pp 198–207

Lumsden J (2008) Handbook of research on user interface design and evaluation for mobile technology. Information Science Reference, Hershey

Lund AM (2008) Measuring usability with the use questionnaire. http://www.stcsig.org/usability/newsletter/0110_measuring_with_use.html. Accessed 17 May 2014

Nielsen CM, Overgaard M, Pedersen MB, Stage J, Stenild S (2006) It's worth the hassle!: the added value of evaluating the usability of mobile systems in the field. NordiCHI '06. ACM, New York, pp 272–280

Nielsen J (1994) Estimating the number of subjects needed for a think aloud test. Int J Hum Comput Stud 41(3):385–397

Ostkamp M, Kray C (2014) Supporting design, prototyping, and evaluation of public display systems. EICS'2014. ACM, New York, pp 263–272

Palen L, Salzman M, Youngs E (2000) Going wireless: behavior and practice of new mobile phone users. In: Proceedings of CSCW '00. ACM, Philadelphia, pp 201–210

Rogers Y (2011) Interaction design gone wild: striving for wild theory. Interactions 18(4):58–62

Roto V, Oulasvirta A, Haikarainen T, Kuorelahti J, Lehmuskallio H, Nyyssönen T (2004) Examining mobile phone use in the wild with Quasi-experimentation. Helsinki Institute for Information Technology, Finland

Schellenbach M, Lövdén M, Verrel J, Krüger A, Lindenberger U (2009) Adult age differences in familiarization to treadmill walking within virtual environments. Gait Posture 31(3):295–299

Singh P, Ha NH, Kuang Z, Olivier P, Kray C, Blythe P, James P (2006) Immersive video as a rapid prototyping and evaluation tool for mobile and ambient applications. In: Proceedings of mobile HCI '06. ACM, New York, p 264

Snowdon C, Kray C (2009) Exploring the use of landmarks for mobile navigation support in natural environments. In: Proceedings of mobile HCI '09. ACM, New York, pp 1–10

van Elzakker CPJM, Delikostidis I, van Oosterom PPJM (2008) Field-based usability evaluation methodology for mobile geo-applications. The Cartographic J 45(2):139–149

van Veen HAHC, Distler HK, Braun SJ, Bülthoff HH (1998) Navigating through a virtual city: using virtual reality technology to study human action and perception. Future Gener Comput Syst 14(3–4):231–242

Winters JM, Story MF, Campbell S, Lemke M, Danturthi D, Barr A, Rempel DM (2001) Mobile usability lab: a tool for studying medical device accessibility for users with diverse abilities. In: Proceedings of the 27th RESNA conference

Challenges of Location-Based Services Market Analysis: Current Market Description

Anahid Basiri, Terry Moore, Chris Hill and Paul Bhatia

Abstract Location-Based Services (LBS) have a huge and rapidly growing market, however LBS market reports do not fully agree about the current market size, the exact number of LBS users, market growth rate, etc. Different market study reports describe LBS market size, number of LBS subscribers and frequent users, revenues and costs and most appreciated application types with different and sometimes extremely contrasting numbers and figures and this becomes more problematic when it comes to forecasting the future market of LBS. One of the very first steps in any market analysis is defining business contribution area, which decides what should be included in or excluded from the market analysis. Location based services has got a large market; therefore it is not an easy task to identify LBS market's boundaries to analyse what is inside the box as LBS applications and exclude the rest. This paper explains the impact of such vague boundaries on market size estimation.

Keywords Location-based services (LBS) · Market report · Business analysis

1 Introduction

Location is clearly one of the most important parameters to describe human behaviour. The rewards for capturing, visualising, analyzing and more importantly using location data to provide more relevant services are huge. The European GNSS Agency (GSA) estimated that 40 % of all applications use location information (GSA Report 2013) and 30 % of all Internet searches are looking for places and locations (W3C 2013). This shows a very promising future and rapid growth for Location-Based Services (LBS). It has forecasted that LBS will have more than double revenue by the end of 2014 in comparison with 2011 (GSA Report 2013).

A. Basiri (✉) · T. Moore · C. Hill · P. Bhatia
The Nottingham Geospatial Institute, The University of Nottingham, Nottingham, UK
e-mail: anahid.basiri@nottingham.ac.uk

G. Gartner and H. Huang (eds.), *Progress in Location-Based Services 2014*,
Lecture Notes in Geoinformation and Cartography,
DOI 10.1007/978-3-319-11879-6_19

Use of location based applications and services, such as navigation and tracking services, location based information retrieval, location based social networking, location based marketing and advertisement, are growing rapidly. The numbers of LBS applications available in the Android and Apple application stores in late 2013 were 700,000 and 775,000, respectively. Comparing theses numbers with 2011s—when the number of available LBS applications in the Android store was merely 88,000—then it can be easily seen that there has been a high rate of growth in the number of LBS application and naturally number of LBS users.

Since LBS is growing with such a high rate, many application developers, business units and even ordinary people are becoming more and more involved with use of location data to develop, provide or received location-based services. As the result, there are many market research teams have been estimating the current market size and forecast LBS future and trends.

LBS market forecast is a challenging task; one should deal with many vague concepts, such as definition of location-based services and its market boundaries, estimation of the impact of incoming features, technologies and social acceptance of them on the market. However some of these parameters, such as privacy and social acceptance of a technology, etc. cannot be easily quantified, therefore describing their impact with figures and numbers can be done in different ways which might result in different future market size. This paper explains the impact of such vagueness, especially vagueness in LBS definition, on its market estimation.

One of the very earliest steps in any market analysis is identifying the area of business contribution. Identifying what type of product/service should be included and what should be excluded is very essential. Similarly for LBS market analysis, the services and applications, which are based on the "location", should be identified firstly. In this regard, LBS definition plays a strong role since it decides what applications and services can be considered as LBS and what should be excluded the market analysis process. However LBS definition, itself, does not have clear boundaries; If LBS is defined as any type of service which uses location data to exclude irrelevant responses (as mentioned in many reports and papers), then number of LBS users will dramatically increase since thanks to GeoLocation API and HTML5, any internet or even off-line mobile user is now provided with LBS as browsers' language settings and mobile phones' time and date is changing according to user's current location; i.e. country. Based on this definition if a mobile phone local time is changed automatically or web browsers' User Interface (UI)'s language is changed based on the country where user is, then almost everyone is an LBS user and a portion of any mobile app revenue should be considered as LBS revenue.

On the other hand, some definitions consider the *essentiality* of location for the service or some definitions exclude services from LBS, if they do not use real-time locations of users to provide service. According to each definition a set of services and applications are included in the LBS market analysis process while it might be excluded by another definition. Consequently, the number of LBS users, size of the revenue and LBS growth rate can be extremely different in different market reports; According to Pew Internet Project report (Pew Report 2013) 74 % of adult

Smartphone owners ages 18 and older use their phone to get Location-based services, such as getting directions or other information based on their current location. While TNS's annual Mobile Life study (2012) estimated only one fifth (19 %) of the world's six billion mobile users are using LBS, with more than three times this number (62 %) aspiring to do so in the future.

Such kind of extreme different, and sometimes conflicting, reports are the consequence of having different definitions for LBS and also related concepts such as location, accuracy, service, etc. behind the scene.

LBS has got a very large market; however due to its definition vagueness, it is not easy to put a bounding box and say what are inside are LBS applications, and the rest is not. Consequently, current market size estimation does not have the same results in different reports and whitepapers. This becomes more differing when it comes to the LBS market forecast.

LBS market has got a broad nature of its applications. Positional data has been a key component of many applications. This makes it difficult to identify a crisp boundary for LBS. Every time it is tried, a new set of applications pops up and suddenly there is revenue that left out of a previous model. Even if the revenue from a new application is apparent, the question arises as to what part of the product or service should really be included in the LBS market (Jacobson 2007).

In summary, vagueness of the LBS-related concepts, boundaries of LBS applications, different possible ways for LBS market segmentation, quantification of impact of social and political views and legislations and many other challenges make LBS market reports and whitepapers very different if not contradictious. Theses are only some examples of such challenges to deal with in LBS market analysis. This paper focuses on impact of having different LBS definitions on evaluation of its market and forecasting its future. Next section describes different views on LBS definitions and related concepts. Section 3 describes impacts of having different LBS definitions on the market size according to different market reports. Finally conclusion and suggestions are provided in the Sect. 4.

2 Location-Based Services Definitions

This section reviews different definitions and views on LBS, related concepts and key components, including location and position, service, positional requirements of LBS, market segmentation, etc.

2.1 Location-Based Services

There have been various definitions for LBS from different perspectives. Koeppel (2000) regards LBS as any service or application that extends spatial information processing, or Geospatial Information system (GIS) capabilities, to end-users via the Internet and/or wireless network. Longley and Maguire (2004) defined LBS as "geographically-oriented data and information services to users across mobile telecommunication networks". Brimicombe and Li (2009) has defined LBS as the delivery of data and information services where the content of those services is tailored to the current or some projected location and context of a mobile user. Voxeo (http://voxeo.com/glossary/what-are-location-based-services/. Access June 2014) defines LBS real-time location intelligence from a customer's mobile device to offer a more personalized experience, or a higher level of service. There are still many other different definitions for location-based services.

Location and service are two essential concepts in LBS definition, which have not been clearly defined and standardised. Location can be assumed as position, which is measurable, or proximity, which can be described numerically or verbally using topological relationships between objects (such as inside), or semantic description, which can help to infer position. Service can be regarded as software or app, hardware, infrastructure and many other possible definitions for "*service*" as are discussed in this subsection.

Position as location still needs to be described by level of precision (resolution). It is not standardised how accurate or precise the position should be, to be able to use it as an LBS input data. Take the Internet browsers' language setting change example; if the Internet browsers (or some mobile apps) have access to (approximate) location of user using mobile communication network, at country level, they can change language, date and local time accordingly. Is this should be considered as an LBS? Is the accuracy of positional data sufficient to call the output "*location-based information*"? It becomes clearer when compare this with another type of service, such as excluding a search results by a search engine according to the city or more accurate location of user where more relevant responses are provided to user according to (more accurate and precise) positional data. Identifying the best level of acceptable accuracy and precision is highly correlated with type of application and service.

The same is also true for temporal accuracy. As it mentioned in some of definitions of LBS, currency of location (or real-time location) can matter (Brimicombe and Li 2009, http://voxeo.com/glossary/what-are-location-based-services/. Access June 2014). Some have defined LBS as services that use "online", "real-time" or "current" location data. However concurrency or temporal accuracy of position can also vary respect to the application. For some applications, such as navigation and tracking real-time data is essential while for some others, such as location based advertising, it is just an option. Other aspects of quality of positional data ISO-19157 (Geographic information—Data quality), such as thematic accuracy (e.g. semantic description clarifying location) can also be considered in this way.

Table 1 shows some of examples of location based services and application by providing some initial requirement for positional data (Basiri et al. 2014). Although there is not only one way for LBS application segmentation, the emphasis of this table is on different positional requirement for LBS application rather than market segmentation.

Table 1 LBS application and positional requirements

Vertical	Application examples	Positioning component's requirement
Navigation and tracking	Navigation	Accuracy of few meters or less
	Positioning	Response in real-time or few seconds (in general applications)
	Path finding	Very high availability (seamlessly available indoors and outdoors)
	Tracking	Very high reliability and continuity
	Asset finding	
Marketing	LB (social) marketing	Accuracy in the order of hundreds of meters
	LB advertisement and LB dealing	Response in few minutes (or even more)
	Proximity-based voucher/offers/ rewards	Medium availability
	Location-based social reward sharing	Medium reliability and continuity
Entertainment	LB social networking	Accuracy in the order of tens of meters (buildings level)
	LB gaming	Response in real-time or few seconds
	LB fun sharing	Medium to high availability (Ideally seamless indoors and outdoors)
	Find your friend	High reliability and continuity
	LB chatting and dating	
Location-based information retrieval	Location-based NEWS	Accuracy from a few meters (for tourist guide and proximity search) to hundreds of meters or even a few of kilometres (for NEWS and weather)
	Location-based Q&A (query)	Response in real-time or few seconds
	Proximity searching	Medium availability
	Tourist guide	High reliability and continuity
	City sightseeing	
	Traffic, weather and transportation Info.	
Safety and security	Emergency services	Accuracy of tens of meters or lower
	Emergency units allocation	Response in real-time or few seconds

(continued)

Table 1 (continued)

Vertical	Application examples	Positioning component's requirement
	Emergency alert services	Very high availability (seamless indoors and outdoors)
	Ambient assisted living	Very high reliability and continuity
	Security surveillance	

Service can be also a vague concept. Considering concepts of *Software as a Service* (SaaS), *Hardware as a Service* (HaaS), *infrastructure as a service* (IaaS), *platform as a servi*ce (PaaS), and *information technology management as a service* (ITMaaS), makes it more complicated since different assumptions result in different figures and numbers for market size. Some of theses concepts include or exclude the cost of infrastructure and platforms or licence of software. In this regard it is very important what is meant by "*service*" and initially to know what part of infrastructure, hardware, software licence, platform, etc. should be included in market size estimation.

In addition to *location* and *service*, there is another resource of uncertainty in LBS definition: the level of dependency of a service to location, i.e. to what extend a service has to depend on positional data to call it an LBS. It is possible to use location to provide an enhanced version of a service while the original version of service can still work. It would be more pleasant for users to have more relevant search results from *Google.com* if their location is attached to their requests, however they could get the original service without such contextual data (i.e. location). The result of service by using location would be more enhanced and user-friendly. Facebook checking-in is another example, user will receive more relevant advertisements and suggestion if he/she checks-into specify his/her current location, however without this, the same service (with less potential relevancy) can be provided. This paper calls these types of services as "location-*enhanced* services".

On the other hand, there are some sorts of services, which cannot work without location, such as navigation services, tracking services, tourist guidance. These services use location (with different level of accuracy) as an essential input data, as without location, the service cannot be valid/provided. This paper calls them "location-*based* services". In this regard, Google search results and Facebook advertisement examples, which are a location-enhanced service, navigation and tracking apps and foursquare are location-based services since location is an essential input for them.

2.2 LBS Market Segmentation

Market segmentation is used to enable a business to better target its products/services at the right customers. It is about identifying the specific needs, wants and desires of customer groups and then using those insights into providing products and services which meet customer needs. Segments are usually measured in terms of sales, types or target and customers' need or volume. In this regard, there are plenty of ways to identify market segments; even taking only one approach it is possible to have overlapping verticals (categories of products or service). For example, Table 1 categorises LBS applications into five categories. In addition to having many approaches to do this, even taking this segmentation example, it is not easy to exclude one application from other four categories. E.g. Table 1 has got location-based social reward sharing in the category of marketing while it can be considered in the category of entertainment, depending on context. This can be a challenge in market size estimation since one application's revenue can be considered twice or totally excluded from market analysis.

There are various methods (or "bases") a business can use to segment a market, some of the most popular are: Geographical, demographical, psychographic and behavioural. One of the most widely-applied approaches to describe the market is geographical distribution of users since location of users are available; as LBS applications or service providers need to know user's location to provide the service. Other can segment the LBS market according to users' age, gender, occupation, etc. One of the best examples of this is dividing the whole market into two groups of military and civil users. (Jacobson 2007) enumerated differences between these approaches of segmentations with others. One also can identify different verticals by considering different types of applications. Therefore, it is possible to generalise two or more types of LBS applications into one or go through details and split one segment into two or more verticals.

Next section provides examples of different market segmentation corresponding different LBS application/service areas, which results in having different estimation for the most growing application types (verticals). This can be confusing for new companies who are taking new policies or targeting new market based on these reports as the most growing segments or numbers of its subscribers can be different according to different reports.

3 Impact of Different LBS Definitions on Market Reports

As it explained in Sect. 2, due to vagueness in concepts of location, service, quality of positional data and dependency of a service and location data, it is possible to define LBS in many different ways. Consequently, from one point of view some location-based applications and services may be excluded from LBS market analysis process, while another view includes them. The impact of vagueness of LBS related concepts, such as location data, accuracy, service, etc., on LBS market is viewed in this section. Different LBS market research groups have considered different definition for LBS and as a result, their reports disagree in regard of market size, numbers of subscribers, most-widely used LBS applications, etc.

Gartner Group forecasts the revenue generated by consumer location-based services to reach $13.5 billion in 2015, of which advertising will be the dominant contributor (Gartner Group Report 2012). Slightly different, Pyramid's research predicts that the global location-based services market revenue to reach US $10.3 billion in 2015, up from $2.8 billion in 2010 (Pyramid Research Report 2014). Meanwhile, Juniper Research expects revenues from mobile location-based services to more than $12.7 billion by end of 2014 (Juniper Research 2014). According to Juniper Research, navigation with maps and GPS is identified as the most popular motivation behind the LBS uptake (46 %), but there is growing interest in more diverse activities, with 13 % of current social network users 'checking-in' through platforms like Foursquare, or Facebook according to this report.

Markets and Markets's recent report (2014) forecasts Location based services market to grow from $8.12 billion in 2014 to $39.87 billion in 2019. This represents a compound annual growth rate (CAGR) of 25.5 % from 2015 to 2019. While according to IT research team Berg Insight Report (2014) the LBS market in North America is forecasted to grow at a compound annual growth rate (CAGR) of 16.1 % from $1.8 billion in 2013 to reach $3.8 billion in 2018, see Fig. 1. This report estimates LBS revenues in Europe to grow from €735 million ($1.01 billion) in 2013 at a CAGR of 25.8 % to reach €2.3 billion ($3.1 billion) in 2018, see Fig. 1. Berg Insight estimates that about 50 % of all mobile subscribers in Europe were frequent users of at least one location-based service at the end of 2013. It also expects the main growth will come from increasing ad revenues in the social networking and local search segments.

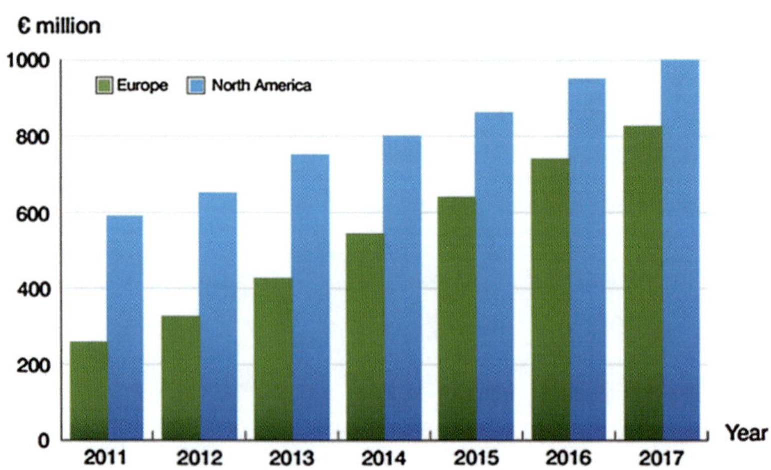

Fig. 1 Mobile LBS revenue forecast by Berg Insight

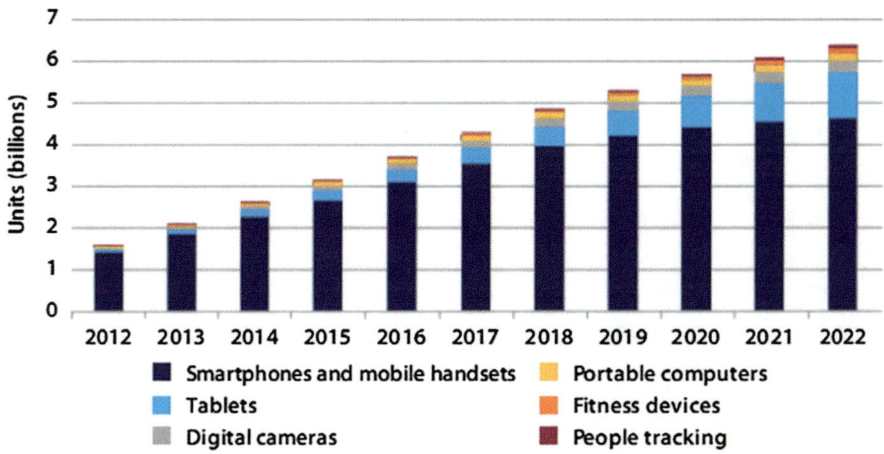

Fig. 2 GNSS-enabled device numbers, by GSA

With a different hypothesis, European Space Agency (GSA Report 2013) considers all GNSS-enable devices shipment as a part of LBS market (Hardware as a Service). It forecasts that only in the EU-27 countries, shipments of GNSS-enabled devices will grow from €218 million to more than €600 million per annum by 2022 while global GNSS-enabled markets are forecast to grow to approximately €250 billion per annum by 2022 and the core revenues (these attributable to GNSS functionality and service sirectly) are expected to reach over €100 billion in the same time, see Fig. 2.

ABI research reports distinguish between location-based services and location enabled services. According to ABI research 2012, global revenue of LBS reaches to 8 Billion dollars and the two most widely used LBS applications are navigation and enterprise, respectively.

As it explained, there are differences in regards of LBS market segmentation, market size, number of subscribers and most appreciated applications and their revenue. It is very important for market report to clarify what they mean by location-based service, what is included and what is not. This make less confusion for policy makers, SMEs, businesspersons who want to take a policy in line with market trends.

In this regard, it is highly suggested that the assumptions and hypothesis are explained in more detailed in all market reports since their economic impact are sometimes more than the size of market. As it explained in last section, this paper suggests distinguishing between Location Based Services and Location Enhanced Services to find the direct/core market size in an easier way.

4 Conclusion

This paper reviewed impact of different LBS definitions and related concepts, such as service, accuracy, and position on its market size, number of users, the most growing market segment and in general on the LBS market. As different assumptions and hypothesis in LBS market analysis result in different figures and numbers to describe market size, it is very important to clarify what is meant by LBS. This help to have a better understanding of what kind of services and application are included in market analysis and what are excluded from. In addition to vagueness in these concepts, another source of ambiguity in market analysis, which causes different (And sometimes conflicting) results for LBS market, is market segmentation approaches and the numbers of LBS verticals have got an impact on the findings of a research report. This paper reviews and evaluated different market reports from these points of views.

Acknowledgements This work was financially supported by the EU FP7 Marie Curie Initial Training Network Multi-technology Positioning Professionals (MULTI-POS) under grant no. 316528.

References

Basiri A, Figueiredo Silva P, Lohan E.S, Peltola P, Hill C, Moore T (2014) Overview of positioning technologies from fitness-to-purpose point of view. In: International conference on localization and GNSS (ICL-GNSS)
Berg Insight (2014) Mobile location-based services, 8th edn. Berg Insight
Brimicombe A, Li C (2009) LBS and geoinformation engineering. Wiley, New York. ISBN 978-0-470-85737-3
Gartner Group Report (2012) Dataquest insight: The top ten consumer mobile applications for 2012
GSA Report (2013) GNSS Market Report, 3rd edn of the European GNSS Agency (GSA), Oct 2013
http://www.w3c.tut.fi/events/2013/0911-techday/slides/W3C_Keynote_GMV.pdf
Juniper Research (2014) Location based services market
Jacobson L (2007) GNSS market and application
Pew Reseach Center (2013) Location-based services, smartphone ownership, Washington DC
Pyramid Research Report (2014) Location based services market

Printed by Printforce, the Netherlands